Contents

1 The computer

Before you start

1 Match the computer parts with the words below.

floppy disk ☐ scanner ☐ mouse ☐ keyboard ☐
tower ☐ monitor ☐ CD-rewriter ☐ printer ☐

Reading

2 Read the text quickly. Match the headings (a–d) with the paragraphs (1–4).

a Memory ☐   c PCs and Notebooks ☐
b Speed ☐   d Hardware/Software ☐

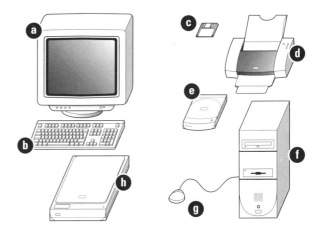

1 The parts of a computer you can touch, such as the monitor or the Central Processing Unit (CPU) are hardware. All hardware except the CPU and the working memory are called peripherals. Computer programs are software. The operating system (OS) is software that controls the hardware. Most computers run the Microsoft Windows OS. MacOS and Linux are other operating systems.

2 The CPU controls how fast the computer processes data, or information. We measure its speed in megahertz (MHz) or gigahertz (GHz). The higher the speed of the CPU, the faster the computer will run. You can type letters and play computer games with a 500 MHz CPU. Watching movies on the Internet needs a faster CPU and a modem.

3 We measure the Random Access Memory (RAM) of the computer in megabytes (MB). RAM controls the performance of the computer when it is working and moves data to and from the CPU. Programs with a lot of graphics need a large RAM to run well. The hard disk stores data and software programs. We measure the size of the hard disk in gigabytes (GB).

4 Computer technology changes fast, but a desktop PC (Personal Computer) usually has a tower, a separate monitor, a keyboard and a mouse. The CPU, modem, CD-ROM and floppy disk drives are usually inside the tower. A notebook is a portable computer with all these components inside one small unit. Notebooks have a screen, not a monitor, and are usually more expensive than desktops with similar specifications.

3 Look at these words from the text. Write H (hardware), P (peripheral), S (software) or M (measurement) next to each one.

1 CPU	H	7 mouse	☐
2 MacOS	S	8 modem	☐
3 megabyte (MB)	M	9 Linux	☐
4 printer	P	10 scanner	☐
5 RAM	☐	11 gigabytes (GB)	☐
6 megahertz (MHz)	☐	12 floppy disk	☐

Vocabulary

4 Match the highlighted words and phrases in the text with the definitions (1–8).

1 parts _____
2 pictures and images _____
3 a way of doing something _____
4 reads and uses data _____
5 measurements _____
6 use a computer program _____
7 keeps data in the memory _____
8 how well a computer does something _____

Speaking

5 Work in pairs. Look at the chart and compare the two computers. Use *fast, slow, cheap, expensive, big, small.*

	Hi-Tech 2010	Series X Wi-Fi
Type	PC	Notebook
CPU	933 MHz	1.5GHz
RAM	256 MB	512 MB
Monitor/Screen	17 inch	15 inch
Hard disk	20 GB	40 GB
Price	€2,000	€2,999

▶ ### Get real

Look at new computers on the Internet or in magazines. Find one you like. Make a note of its specifications. Bring your notes to class and say why you like it. Build a class file of computers with information about them.

2 The desktop

Before you start

1 What do you see first when you turn on a computer? How do you open a program?

Reading

2 Read the text quickly and match the headings (a–d) with the paragraphs (1–4).

a The control panel ☐   c The desktop ☐
b The drives ☐   d Using icons ☐

1 The desktop is the screen that appears after you boot up, or turn on, your computer. It shows a number of icons on a background picture or colour. When you buy a new computer and boot up for the first time, the desktop will only show a small number of icons. In the Windows operating system, these usually include My Computer and the Recycle Bin.

2 Double-clicking on an icon with the mouse opens a computer program, a folder or a file. Folders usually contain other files. You can move icons around the desktop, add new ones or remove them by deleting them. Deleted files go to the Recycle Bin. People usually put the programs they use most often on the desktop to find them quickly.

3 When you double-click on My Computer another screen appears. This screen shows the A: drive icon, for floppy disks; the C: drive icon, which usually contains all of the main programs and folders on your computer; the D: drive icon, which is usually the CD-ROM drive, and the Control Panel folder.

4 When you double-click on Control Panel, another screen appears that shows many other icons, such as the Display icon and the Date/Time icon. Double-clicking on Display opens a box that lets you personalize your desktop by changing the screen saver (the moving image that appears when no one is using the computer) or the background picture.

3 Decide if the sentences are true (T) or false (F).

1 The desktop appears before you boot up. T/F
2 Files are usually inside folders. T/F
3 People usually put their favourite programs on the desktop. T/F
4 Use the C: drive to open floppy disks. T/F
5 You cannot change the background picture of the desktop. T/F
6 The Control Panel folder contains the Date/Time icon. T/F

Vocabulary

4 Find the words in the text that mean:

1 comes into view so you can see it (paragraph 1) _____

2 the picture or colour on your screen (1) _____

3 clicking the mouse two times quickly (2) _____

4 something that holds documents or files (2) _____

5 most important (3) _____
6 make something the way you want it (4) _____

5 Complete the sentences (1–7) with the words in the box.

> Display ■ screen saver ■ folders ■ Recycle Bin
> ■ files ■ deleted ■ desktop

1 The _____ icon lets you change the way your desktop looks.
2 If you remove a file by mistake, you can find it in the _____.
3 The _____ appears when you don't use the mouse or keyboard.
4 I didn't use that program very much so I _____ it from my desktop.
5 I have a great program on my _____ that I use for playing music.
6 Windows Explorer lets you move _____ from one folder to another.
7 _____ contain documents or files.

Speaking

6 Choose five icons on your desktop. Say what you use these programs for.

▶ ***Get real***
Go into Control Panel on your computer and choose two other icons that interest you. Double-click on them and make notes on what they do. Report back to the class.

3 Using a word processor

Before you start

1 Look at the notebook keyboard below. Answer the questions.

1 Which key is between G and J? _____
2 Which key is to the left of Y? _____
3 Which key lets you type in capital letters? _____
4 Where are the arrow keys? _____
5 Where is the multiplication sign? _____

2 Work in pairs. Choose a letter or key from the keyboard and describe where it is. Do not say which key you have chosen. Use *next to, above, below, between, on the right/left/top/bottom.*

3 Discuss these questions.

1 How often do you type letters or documents?
2 **Which** word-processing program do you use?
3 Which *commands* do you know?
4 How many different **fonts does** this **question** have?
5 Find the words in 1–3 that are in bold, in italics, underlined and highlighted.

Reading

4 Look at the table of word-processing tools and commands and their functions. Then answer the questions.

1 Which tool checks your spelling? _____
2 Which edit command removes text or images? _____
3 Which format command changes the letter size? _____
4 Which file command opens a file or document? _____
5 Which insert command lets you put in an image? _____
6 Which view command shows the document as a printed page? _____

Close	This command closes the open file.
Copy	This editing command copies any highlighted text or images and keeps it in memory. We say anything copied is on the clipboard.
Cut	This editing command deletes any highlighted text or image.
Font	Formats, or changes, the type style and size of the characters.
Full Screen	This view command makes the open document cover all of the screen. This also hides the menu bar and the toolbar so that you cannot see them. The menu bar shows commands and tools in words; the toolbar shows them with icons.
Language	This tool opens a thesaurus to help you find synonyms and antonyms (similar and opposite words).
Open	Opens a file from one of the computer's drives.
Paragraph	Formats the paragraph settings to change the way the paragraph looks.
Paste	This editing command puts anything that is on the clipboard onto the screen.
Picture	Inserts a picture or image into your document.
Print Layout	Views the open document as it will look when it is printed.
Save	Saves the open file or document.
Save As	Saves the file to another location, with another name or in a different format.
Spelling and Grammar	This tool checks the document or any highlighted text for spelling and grammar errors.
Symbol	You can insert many special characters with this command.
Undo	This editing tool cancels the last command. It does not work with every command.

5 Write the commands and tools from the table under the correct heading.

File	Edit	View
_____	_____	_____
_____	_____	_____
_____	_____	
_____		

Insert	Format	Tools
_____	_____	_____
_____	_____	_____

6 Match the first part of the sentence (1–6) with the second part (a–f).

1 Typing letters with a word processor
2 Many companies need people
3 I can learn a lot of new words
4 People usually type business letters
5 Check your spelling and grammar
6 If you cut a sentence out by mistake,

a in the Times New Roman font.
b by using the thesaurus.
c before you print out your document.
d who can use a word processor.
e try clicking the undo button.
f is easier and quicker than writing by hand.

Vocabulary

7 Complete the sentences with words from the box.

> character ■ clipboard ■ fonts ■ format
> locations ■ menu bar ■ settings ■ toolbar

1 When you copy text, it stays on the _____ until you want to paste it.
2 Change the paragraph _____ if you want bigger spaces between the lines.
3 Each word on the _____ contains a list of commands and tools.
4 Most of the icons on the _____ are also in words in the menu bar.
5 You can save a file in many different _____ in your computer.
6 MS Word has about a hundred different _____ for you to choose from.
7 A word processor lets you _____ a paragraph as well as the font.
8 If the _____ you want is not on the keyboard, look in the Symbol command.

Speaking

8 Match the icons (a–l) with the words (1–12). Say what the command or tool does.

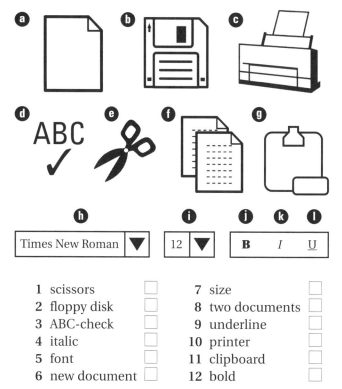

1 scissors ☐
2 floppy disk ☐
3 ABC-check ☐
4 italic ☐
5 font ☐
6 new document ☐
7 size ☐
8 two documents ☐
9 underline ☐
10 printer ☐
11 clipboard ☐
12 bold ☐

A *What's f?*
B *It's a picture of two documents.*
A *What does it do?*
B *It copies text or images onto the clipboard. / It lets you copy text or images onto the clipboard.*

Writing

9 Practise your typing! Type a paragraph of any English text that is new to you using a word-processing program. Do the following:

• format the verbs in bold
• put nouns in italics
• underline any adjectives
• use the thesaurus to check any words you do not know
• change the font and the font size for each sentence
• use the spell check to check your work.

Then print your document.

> ▶ *Get real*
> Go to the menu bar and look at File, Edit, View, Insert, Format and Tools. Find out the function of two other commands or tools and use them in your document from Exercise 9. Report back to the class and make a class file of the new commands and tools.

Before you start

1 Work in pairs and discuss the questions.

1 Do you like writing by hand? Why?/Why not?
2 Have you ever used a typewriter or word processor? Which word processor?
3 What are the differences between handwriting, typing and word processing?

Reading

2 Read the text and underline the advantages of word processing in the first paragraph and the disadvantages in the second. Write the number of each.

Advantages ☐   Disadvantages ☐

The case for and against
WORD PROCESSING

People use word processors for writing all kinds of documents, such as letters, school papers and reports. Word processors have many advantages over handwriting and manual typewriters. Word processing is faster and easier than writing by hand and you can store documents on your computer, which you cannot usually do on a typewriter. This makes it easier to review and rewrite your documents. You have more formatting choices with a word processor, and the spelling, grammar and language tools are useful, too. You can also print copies of your documents, which look neater than handwritten ones. Many language students use word processors to improve their writing skills and because they help them feel proud of their work.

Word processors do have disadvantages, however. First, it is not easy to read long documents on a computer screen. Second, sometimes the printer does not print an exact copy of what you see on the screen. Not all word processors can read each other's files, which is another disadvantage. Finally, word processors do not always work well with e-mail. If you paste a word-processed letter into an e-mail it may lose a lot of its formatting. Many people use a text editor for the Internet, which is similar to a word processor but has fewer formatting features and cannot use graphics. Text editors, such as Notepad, use a simple coding system called ASCII (American Standard Code for Information Interchange), as does e-mail.

3 Decide if the sentences are true (T) or false (F).

1 You can store letters on a manual typewriter.   T/F
2 You can change your documents easily on a word processor.   T/F
3 Printed documents look better than handwriting.   T/F
4 Improving your writing is more difficult with a word processor.   T/F
5 Word processors work well with e-mail.   T/F

Vocabulary

4 Match the highlighted words and phrases in the text with the definitions (1–6).

1 by hand, not electronic _____
2 the way a program organizes data _____
3 a program used for simple text files _____
4 the code that e-mail uses _____
5 things that a program has, or can do _____
6 a program used for text and graphics _____

Speaking

5 Work in groups. Which of these documents would you write by hand and which on a word processor? Say why.

> a formal letter ▪ an informal letter
> ▪ an invitation to a party ▪ a birthday card
> ▪ a shopping list ▪ an application form
> ▪ a note to your teacher/friend/father
> ▪ a school report ▪ a 'for sale' notice

Writing

6 Write a short paragraph about some of the advantages of writing with pen and paper. Use the following ideas to help you. Add any other ideas you may have.

• pen and paper – cheap
• you can write anywhere (don't need electricity)
• don't need to learn to type
• handwritten letters – friendlier & more personal

▶ Get real

Take your paragraph from Exercise 6. Type it into the word processor on the computer you use. Change or add some formatting features, such as the font, bold, italic or underline. Copy and paste the formatted letter into a text editor such as Notepad. Report back to the class on which formatted features did not appear.

5 | Storing data

Before you start

1 What information can you store on a computer? Where can you store your documents or files?

Reading

2 Read the text quickly. Write the number of the paragraph that tells you about:

a computer directories ☐
b organizing personal files ☐
c where you can store data ☐

1 Computers store letters, reports, pictures, music and video as data. You can store data on floppy disks but they are too small to hold most pictures or videos. You can store these on a CD-ROM but you need a CD-re-writer to copy the data. People often use portable removable disks, which can store up to one gigabyte of data and do not need any expensive hardware. Most computer owners store their data on the hard disk but because computers can crash, they often use other disks to make back-up copies.

2 Computers store program files on the hard disk, which is usually the C: drive in the Windows OS (operating system) or the Macintosh Hard Disk in Apple computers. Computers store program files in folders and organize these folders in a directory (see below). The plus sign (+) means that the folder contains other folders or files. Clicking the plus sign next to the icon opens the other folders and files in it. Clicking the minus (–) sign closes the folder.

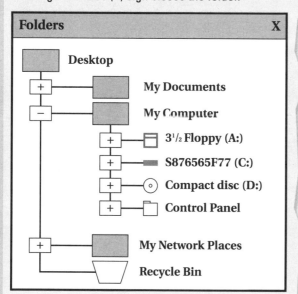

| Folders | X |

Desktop
+ My Documents
– My Computer
 + 3¹/₂ Floppy (A:)
 + S876565F77 (C:)
 + Compact disc (D:)
 + Control Panel
+ My Network Places
Recycle Bin

3 It is important to organize your files. Windows has a folder called My Documents to help you do this. It contains other folders called My Music, My eBooks, My Videos and My Pictures. Saving files on your hard disk without any order makes them difficult to find.

3 Write short answers to these questions.

1 Name four ways you can store computer data.
a _____ c _____
b _____ d _____

2 Name the open folder in the diagram.
_____

3 How many folders are in My Computer in the diagram? _____

4 What is the hard disk called in Apple Macintosh computers? _____

5 Where will you find My Videos in the Windows OS? _____

4 Match the questions (1–5) with the answers (a–e).

1 Why can't a floppy disk store videos?
2 What are removable disks?
3 Where do computers store program files?
4 Why do you need to organize your files?
5 What happens if you click the plus sign?

a Portable data storage disks.
b Other folders or files appear.
c So that you can find them easily.
d On the hard disk.
e Because it can't hold a lot of data.

Vocabulary

5 Find the words in the text that mean:

1 computer information (paragraph 1) _____
2 when computers stop working (1) _____
3 a second copy of a file (1) _____
4 to put in neat order (2) _____
5 a list (2) _____

Speaking

6 Work in pairs. Look at the files below. How would you organize them so that you can find them easily? Add more files to the list.

music files ■ history essays ■ games
■ pictures of my friends ■ videos
■ photos of my family ■ my science reports
■ letters to my friends ■ letters to my family
■ my English language lessons
■ desktop pictures ■ screen savers

▶ ## Get real

Ask people in your family or your friends' families who have their own computers how they organize their personal files. Ask them what categories they use and how they decide what files go into each category. Make a class list of the types of data they store.

6 | Creating a folder

Before you start

1 Match the icons (a–i) with the words (1–9) below.

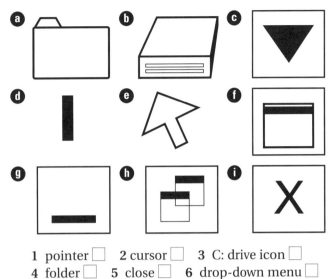

1 pointer ☐ 2 cursor ☐ 3 C: drive icon ☐
4 folder ☐   5 close ☐   6 drop-down menu ☐
7 minimize ☐ 8 maximize ☐ 9 restore ☐

Reading

2 Read the text quickly. What do you use from Exercise 1 to make a folder?

1 To make a new folder in the Windows OS, go to the Desktop, find the My Computer icon with the pointer and double-click it using the left mouse button. The My Computer window appears, showing the different drives. Maximize the screen if necessary.

2 Double-click the C: drive icon. The C: drive window appears showing the folders in your C: drive, either in a row or in a list.

3 Move the pointer to the menu bar. Click on File and a drop-down menu appears. You can only click the words New or Close.

4 Move the pointer to the word New. Another menu appears with Folder at the top of the list.

5 Click on Folder. This creates a new folder that appears at the end of the list of folders on the C: drive. The words New Folder are highlighted. The cursor also flashes on and off to show you where to type.

6 Click on New Folder and type the name you want in the box. This can be up to 250 characters long, but you cannot use the characters '\ / : * ? " < > |' in your folder name.

7 Click anywhere on the window to see your new folder name. If you do not click on the window, you will save your new folder as New Folder, not with the name you want.

8 Close the window.

9 Your new folder is now listed in the C: drive in alphabetical order.

3 Match the diagrams (a–d) with the instruction numbers from Exercise 2.

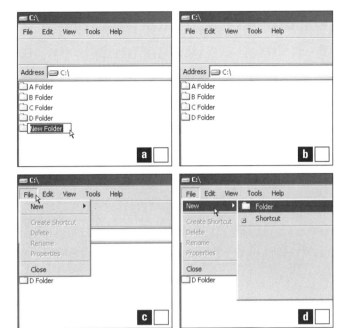

Vocabulary

4 Complete the sentences with words from Exercise 1.

1 Double-click the _____ to view a list of folders and files.

2 You will find the Undo command in the Edit _____ menu.

3 The _____ shows you where to type on the screen.

4 The mouse controls both the _____ and the cursor.

5 If you click _____, the window will cover all of the computer screen.

6 Clicking _____ changes the size and location of the window.

Writing

5 Write down the instructions you need to operate one of the following:

> a mobile phone ■ a tape recorder ■ a camera ■ a video recorder ■ a CD/cassette player

▶ Get real

Imagine that your class is going to store all the information from your English lessons on computer, so that any student can use it. How could you organize the information into folders (e.g. grammar) and files (e.g. the present simple)? Create a list of folders and files, and name them all.

Before you start

1 How is it possible to lose files on a computer?

Reading

2 Read the text quickly and match the headings (a–e) with the paragraphs (1–5).

a Saving existing files ☐   c Saving new files ☐
b Naming files ☐   d Defaults ☐

① Programs that let you create files or save data have a Save command, usually in the File menu. When you save a new file, the Save As dialog box appears (see below). You can let the computer decide the location, the file name and the format, or you can choose these settings yourself. There are many different file formats and they all have advantages and disadvantages. You can save a word-processing document as a web page, for example, or you can save **digital** photographs in a JPEG format, a TIFF format or many others.

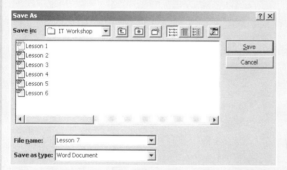

② The Save As command is the **default** command for any new document so the Save As dialog box appears even if you click Save. If you do not give a new document a file name in a word-processing program, the file name will usually default to the first line in the document. The default file name in graphics programs is usually '**Untitled**'.

③ If you work on an **existing** file and want to save changes, click Save, not Save As. You can use Save As to save an existing file in a different location, for example, in another drive or folder (using **Save in**), with a new name (using File name) or in a different format (using **Save as type**).

④ Having different folders helps you organize your files, but naming all the different files in one folder is not always easy. You should give files names that describe exactly what they contain so that you can find them easily.

3 Decide if the sentences are true (T) or false (F).

1 If you do not choose a location, a name, or a format when you save a new file, the computer will do it for you.   T/F
2 If you click Save on a new file, the Save As dialog box appears.   T/F
3 You can save files in one format only.   T/F
4 You can save existing files in more than one place on your computer.   T/F
5 The Save command only saves existing documents.   T/F
6 Most word-processing documents use 'Untitled' as the default name.   T/F
7 The Save As command cannot save existing documents.   T/F
8 Thinking of names for your files is easy.   T/F

Vocabulary

4 Complete the sentences (1–6) with the **highlighted** words and phrases in the text.

1 You can make back-up copies of _____ files on a floppy disk by using Save As.
2 I need a _____ camera because I want to save my pictures on my computer.
3 Graphics programs save files without names as _____ by default.
4 Clicking _____ will let you go to the A: drive, C: drive or D: drive.
5 Programs have _____ settings for all kinds of functions.
6 Click _____ in the Save As dialog box to change the file format.

Speaking

5 Work in pairs. Bring to your class six photographs of your family, friends, city or pets that you would like to store on your computer. Decide on file names to give them. Use the most important part of the photograph, but try to keep the names short.

I'd call this file 'Dad Sleeping 50', because I took the picture on his 50th birthday and he's sleeping in a chair.
I wouldn't call this 'Picture 27' because the file name doesn't describe the picture.

▶ Get real

Type *Photographs* in your search engine. Visit at least three websites that contain photographs or other visual images. Look at the categories that the sites use to group their photographs. Report back to class to say what each category contains. Write up the category list for your next class.

8 The Internet

Before you start

1 Have you ever surfed the Internet? Which websites did you visit?

Reading

2 Read the article. Decide if the sentences are true (T) or false (F).

1 The Internet first started in the USA. T/F
2 The Internet and the WWW are different. T/F
3 Berners-Lee invented the Internet. T/F
4 One file on the WWW can have two or more addresses. T/F
5 There are 40 million Internet users today. T/F

THE INTERNET originated in the early 1970s when the United States wanted to make sure that people could communicate after a nuclear war. This needed a free and independent communication network without a centre and it led to a **network** of computers that could send each other e-mail through **cyberspace**.

Tim Berners-Lee invented the World Wide Web (WWW) when he discovered a way to jump to different files on his computer using the **random**, or unplanned, links between them. He then wrote a simple coding system, called HTML (Hyper Text Markup Language), to **create** links to files on any computer connected to the network. This was possible because each file had an individual address, or URL (Uniform Resource Locator). He then used a set of **transfer** rules, called HTTP (Hyper Text Transfer Protocol), to link Web files together across the Internet. Berners-Lee also invented the world's first browser. This lets you locate and view Web pages and also **navigate** from one link to another.

The WWW became available to everyone in 1991 and the number of Internet users grew from 600,000 to 40 million in five years. Today, that number is much larger and there are now many browsers that provide Web pages, information and other services. You can also do research, download music files, play **interactive** games, shop, talk in chat rooms and send and receive e-mail on the WWW.

3 Find the correct word or abbreviation in the text.

1 an address for Web pages _____
2 a coding system that creates links _____
3 this finds and shows Web pages _____
4 rules for transferring files _____
5 a group of computers joined together _____

Vocabulary

4 Match the groups of verbs below with their general meaning from the box.

move ■ make, start ■ join ■ look at ■ find

1 browse, surf, view _____
2 download, navigate, transfer _____
3 connect, link _____
4 discover, locate _____
5 originate, create, invent _____

5 Complete the sentences (1–7) with the highlighted words from the text.

1 Some people spend too much time playing _____ games on the Internet.
2 You can sometimes have a computer _____ that is not connected to the Internet.
3 It is easy to _____ around a screen with a mouse.
4 Berners-Lee discovered how to _____ links between computers in new ways.
5 Some people surf the Internet at _____, just to see what they can find.
6 People use the Internet to _____ information from one place to another.
7 When you surf the Internet, you are travelling in _____.

Speaking

6 Work in groups. Say which of the following ideas about the Internet are good or bad.

independent ■ world wide ■ cheap to use ■ expensive to buy computers ■ the information may not be true or correct ■ spend too much time playing games ■ talking in chat rooms ■ make new friends ■ visit many interesting websites ■ wait for a long time to download Web pages

▶ **Get real**

Use a browser to surf the Internet at random. Find five interesting websites to tell the class about. Write down the URL of each website and bring the list to class. Build a class file of interesting sites so that other students can visit them.

9 Research on the Internet

Before you start

1 Where is the best place to find information on these topics?

- European history
- the price of mobile phones
- your favourite pop star

2 What are the advantages and disadvantages of finding information from these sources?

> books ■ magazines ■ newspapers ■ libraries ■ encyclopaedias ■ friends or family ■ teachers ■ CD-ROMs ■ television ■ radio ■ the Internet

Reading

3 Read the text quickly and choose the correct answer.

1 Google is a *keyword / search engine*.
2 This WORD is in *upper / lower* case.
3 *AND / WHEN* is a logical operator.

Finding information on the World Wide Web needs an Internet search engine such as Google, Alta Vista or Excite. Search engines have a text box where you type in a keyword or words. A search engine is a software program that reads the keywords in the text box and searches the Internet for Web pages, websites and other Internet files that use them. These documents are shown on the computer screen in a results listing.

When carrying out searches, you should usually be specific and brief in your choice of words. If the keyword is too general, or includes too many different meanings, the results listing may not be useful. Different search engines categorize information in different ways, which changes the way they store and retrieve it.

Using upper case letters (capital letters) in a keyword search will only retrieve documents that use upper case. Typing in lower case (no capitals) is usually better because search engines will retrieve documents that use both upper case and lower case letters.

You can narrow a search using logical operators such as AND, OR and NOT. AND retrieves all the words typed in the text box, OR retrieves either of the words and NOT excludes words. Spelling is important when typing in keywords, but a search engine will not usually read punctuation, prepositions and articles.

4 Tick (✓) the good things to do to find information on the Internet.

1 Choose keywords that are different to the item you want. ☐
2 Give the best keyword to describe what you want. ☐
3 Use as many general keywords as possible. ☐
4 Try to use a keyword that can have only one meaning. ☐
5 Type your keywords in lower case only. ☐
6 Use logical operators to narrow your search. ☐
7 Use full stops and commas. ☐
8 Do not use words like *at, in, on, a/an* and *the*. ☐

Vocabulary

5 Find the words and phrases in the text that mean:

1 clear and exact (paragraph 2) _____
2 put into similar groups (2) _____
3 to bring back (2) _____
4 make smaller (4) _____
5 mathematical words (4) _____
6 does not use (4) _____

6 Are the words in the groups below listed from general to specific or specific to general? Write G→S or S→G.

1 telephone → mobile phone → Nokia _____
2 mother → family → humans _____
3 writing → essay → sentence _____
4 Big Ben → London → UK _____
5 cars → vehicles → transport _____
6 cars → German cars → BMW _____

Speaking

7 Talk about the keywords you should use to find information on the following topics.

- information on cheap hotels in the UK
- what the weather will be like tomorrow
- an essay on the history of the European Economic Union
- mobile phones that connect to the Internet
- a nice present for your mother's birthday.

> ▶ **Get real**
> Carry out the searches in Exercise 7. Then:
> - note the words you used in the search
> - note the top five results for each search
> - visit each site and find out if it is useful.
> Do the search using a different search engine. Bring the list of keywords and your notes on the search results back to class.

Before you start

1 Compare how many e-mails, phone calls and letters you make/send and receive each week.

Reading

2 Read the different opinions. Which one do you most agree with?

Which do our readers like using most: e-mail, telephones or the post? Here are three typical responses from last week's survey.

Lida, 28

For me it has to be e-mail. It's very fast, cheap and modern – you can download music and video, send letters and pictures, and it's informal, which I like. I know privacy and security can be problems but who sends important documents by e-mail? I get annoyed if I get hundreds of e-mails at work and they all expect an instant response, and obviously I hate getting spam, or even worse, a virus.

Jarek, 65

Well, I use all three, but I prefer the phone. It's more expensive, especially for long-distance calls, but I like the instant interaction and I think you can understand more when you hear a person's voice. I like the informality and speed and you can also use your mobile phone for e-mail and sending images. With mobile phones you don't get a lot of unwanted communication, apart from the occasional wrong number.

Andrea, 39

I like modern things, but I still prefer the post. I know postal delivery is slow, but it's cheap, and you can be sure no one will read your mail or listen to your conversations. You can send anything by post, which you can't do with e-mail. Personally, I like receiving handwritten letters – they look, feel and smell different from e-mails. I think it's sad that young people don't write letters now – they're usually more formal than e-mail and students can practise their grammar and spelling. Now, what I don't like is getting is bills and junk mail!

3 Read the quotes again. Tick (✓) the features of each type of communication.

	E-mail	Telephones	Post
cheap	☐	☐	☐
send pictures/images	☐	☐	☐
instant delivery	☐	☐	☐
instant reply	☐	☐	☐
interactive	☐	☐	☐
modern	☐	☐	☐
private	☐	☐	☐
secure	☐	☐	☐
slow	☐	☐	☐
send sound	☐	☐	☐
unwanted communication	☐	☐	☐
usually formal	☐	☐	☐

Vocabulary

4 Which of the words in the box do people usually think of as positive? Which do they think of as negative?

bills ■ communication ■ instant ■ interaction ■ privacy ■ security ■ spam ■ virus

Speaking

5 Work in pairs. How do you feel about getting these unwelcome messages? Why? Add other types of unwanted communication to the list.

wrong number phone calls ■ spam ■ viruses ■ junk mail ■ joke calls ■ bills ■ calls from telephone salespeople

I don't mind getting …
I don't really like getting …
I really don't like / can't stand / (really) hate …

Writing

6 Write a paragraph describing the advantages and disadvantages of e-mail or telephones or the postal service.

▶ **Get real**

Work in pairs. Send each other a handwritten letter in English through the post. Also send each other an e-mail. (They can have the same content.) Describe how the letter looked, how it felt and how it smelt when you received it. Compare the letter to the e-mail you have received. Tell the class which you preferred and why.

Before you start

1 Answer the questions. Then discuss in pairs.

1 Do you use a mobile phone?
2 What do you use it for? Make a list.
3 When is it a good or bad time to make/receive mobile phone calls?

Reading

2 Label the parts of the mobile phones with the words in the box.

> antenna ■ flip cover ■ display screen ■ faceplate ■ keypad ■ scroll keys

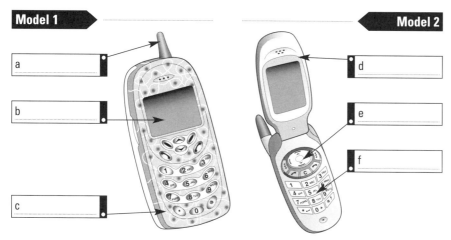

| Model 1 | Model 2 |

a

b

c

d

e

f

3 Read the adverts to check your answers to Exercise 2. Which phone is best for a business person and which is best for a student? Then write the correct names by Model 1 and Model 2 above.

The NEW Brightman QT1

This new super-cool model gives you the best in mobile phones
● multimedia messaging to send snapshots to your friends
● downloadable games, graphics and music
● infrared signal transmission to your computer (no cables!)
● voice-activated dialling – just speak to call
● programmable ring tones so you know who is calling before you answer
● detachable faceplate so you can change it to match your clothes
● antenna for clearer reception

SPECIFICATIONS
BAND MODE digital
TALK TIME 4 hours
STANDBY 5 days
DIMENSIONS 8 x 4 x 1cm

The SP5 Deluxe

Everything you need in mobile technology!
- multimedia messaging with pictures and video
- make calls while browsing the Web
- digital and analog band modes for town and country
- voicemail to send messages to your office
- large screen with 6 text lines for text messages
- large key pad and scroll keys for easy navigation
- Personal Information Manager (PIM) for your appointment schedule
- wireless connection to your PC and headset

SPECIFICATIONS

Band mode	dual
Talk time	5 hours
Standby	7 days
Dimensions	5.5 x 4 x 1cm

4 Which mobile phone has these features? Write *QT1*, *SP5* or *Both*.

1 can work anywhere _____
2 a diary _____
3 a camera _____
4 no cables _____
5 faceplates you can change _____
6 a one-week standby time _____
7 ring tones you can program _____
8 games you can download _____

Vocabulary

5 Which of the words in the box are specific to phones/IT and which are used in general English?

> band mode ■ connection ■ detachable ■ dimensions ■ dual ■ navigation ■ ring tones ■ text messages ■ transmission ■ voice mail

Speaking

6 Work in pairs. Text messages or SMS (Short Message Service) use abbreviations. Match the text messages (1–5) with their meanings.

1 gtg 2 brb 3 thx 4 J4F
5 I k%d meet u @ 7

> I could meet you at 7.00. ■ Be right back. ■ Thank you. ■ Got to go. ■ Just for fun.

Writing

7 Write a paragraph about the mobile phone you have or the one you would like to have.

▶ **Get real**
Use the Internet or magazines to find a new, up-to-date mobile phone. Make a list of the features it has and report back to the class.

Before you start

1 Which of these things do you do with e-mail and which do you do with letters? Compare your answers with another student.

> write a subject ■ send copies ■ write an address ■ add attachments ■ sign in ■ sign your name ■ go to your inbox ■ use a post box ■ click on a name

2 E-mails, like letters, should have a start and an end. Which phrases usually start a message and which end one? Write S (start) or E (end) next to the phrase.

1 Yours sincerely, _____
2 Love and kisses to all. _____
3 Dear Sir or Madam, _____
4 Thanks for your e-mail. _____
5 Give my regards to your family. _____
6 Good to hear from you. _____

Which are formal (F)? Which are informal (I)? Write F or I.

3 Write these messages in the correct order. Which are formal and which are informal?

1 e-mail 21st your August. I to refer dated
_____

2 your I e-mail thanks. yesterday, got
_____

3 you. can't I see wait to
_____

4 seeing look to you. forward I
_____

5 me a Give if you need ring me.
_____

6 require call if assistance. Please you
_____

Reading/Speaking

4 Work in pairs, A and B. Each of you has a box of commands and fields and a diagram of a typical e-mail Compose window with some of the commands and fields missing. You also have information about the commands and fields in your diagram.

- Look at your tables and diagrams before you start the activity.
- Take it in turns to ask and answer questions about your missing commands and fields.
- Write them in the spaces on your diagram.
  A: Start at the top of the next column.
  B: Start on page 15.

Student A

Find out where to write the command or field in the box by asking questions like these:

Where is the Compose command? What does it do? Where is the To: field? What do I type?

> **Command:** Compose   Sign Out   Help   Send   Add/Edit Attachments   Contacts
> **Field:** To: Subject:

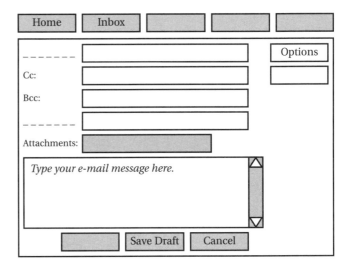

Now use your table to answer Student B's questions. Give answers like these:

The Home command is the first one top left. It takes you back to …
The Bcc: field is the third one. You type …

Command /Field	Information
Bcc: (*blind carbon copy*)	Type an address in this field to send a Bcc:. The person you send the e-mail to does not know who gets a Bcc:. You do not have to send a Bcc: – it is optional.
Cancel	Stops the computer sending the e-mail.
Cc: (*carbon copy*)	Type the address of the person you want to send a copy to. To send multiple copies, type in other e-mail addresses. Separate the addresses with a comma (,). This field is optional.
Home	Takes you back to the opening page of the e-mail program.
Inbox	Takes you to your inbox where you can see your list of messages.
Options	Gives you a number of choices about your e-mail, e.g. changing your password or stopping spam.
Save Draft	Opens your Draft folder to save an unfinished e-mail.

Student B

Student A will ask you questions about his/her missing commands and fields. Use the table below to give answers like these:

The Compose command is the third box. It gives you a new screen ...

The To: field is the top field. You type ...

Command /Field	Information
Add/Edit Attachments	Opens a window so you can attach files to your e-mail.
Compose	Gives you a new Compose screen.
Contacts	Gives you a list of the e-mail addresses in your e-mail program.
Help	Opens a Help screen that gives information on writing a message.
Send	Sends your e-mail message to the recipient.
Sign Out	Closes your e-mail program.
Subject:	Type the topic of the e-mail in this field. This field is usually optional so you can leave it empty.
To:	Type the address of the person you are sending the e-mail to (the recipient) in this field. If you use the Contacts list, you can just click on a name.

Now complete your diagram by asking Student A about the missing commands and fields in your table. Use questions like these:

Where is the Home command? What does it do?

Where is the Bcc: field? What do I type?

Command: Home   Inbox   Options
Save Draft   Cancel
Field. Bcc.   Cc.

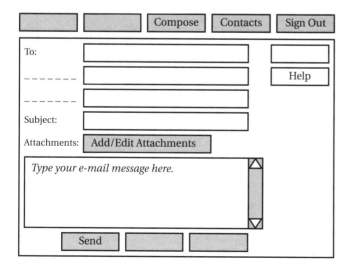

5 Match the questions (1–5) with the answers (a–e).

1 I want to send Katrina a copy of my e-mail to Petr, but I don't want Petr to know I sent her a copy. What do I do?

2 Can I invite all of my friends to my party by e-mail?

3 I got a message telling me the Subject Field is empty. Do I have to type anything?

4 Petr saw me type in my password. Now I'm worried he can read my e-mail. What can I do?

5 I'd like to send a photograph of my birthday party to grandpa. What do I do?

a Sure. You can send multiple copies by typing all the e-mail addresses in the Cc: field.

b Click on the Add/Edit Attachments command and attach the file to the e-mail.

c No. That field is optional in your e-mail program. You don't have to complete it.

d Use Bcc:. This field means that the recipient doesn't know who has received a copy.

e Click on the Options command and change it from there.

Vocabulary

6 Match the words and phrases (1–9) from the texts with the definitions (a–i).

1	recipient	a	a copy of your e-mail
2	field	b	an unfinished document, e.g. an e-mail to send later
3	multiple	c	a secret word
4	password	d	the topic or heading of an e-mail
5	draft	e	something you can choose to do or not do
6	optional	f	the person who receives the e-mail message
7	carbon copy	g	with nothing inside
8	subject	h	a text box where you type words or data
9	empty	i	more than one

Writing

7 Go to your e-mail program and send an e-mail about your English lessons to some of your friends using Cc:. Think of a title or heading for your e-mail and type it in the subject field. Send a Bcc: to your teacher.

▶ ## Get real

Go to the Options window in your e-mail program and choose some features to change on your e-mail. Report back to class on the changes you made and why you made them.

13 E-mail addresses and servers

Before you start

1 Work in pairs. Write down three or four e-mail addresses that you know. What do the different parts of the address mean?

Reading

2 Read the text quickly. Which paragraph (1–4) gives you the answers to the questions (a–d)?

- a What do the two types of mail server do? ☐
- b What are the parts of an e-mail address? ☐
- c How many types of e-mail client are there? ☐
- d What happens when you send an e-mail? ☐

Introducing e-mail addresses and servers

New ▾ | Reply | | | | Send & Receive ▾ | |

1 An Internet e-mail address has a user name, the *at* symbol (@), and a domain name. The user name is the name you choose. The domain has two parts separated by a dot (.). The first part is the network that receives the e-mail and the second is the top-level domain (TLD), which shows the type of organization, such as commercial (.com) or educational (.edu). Sometimes the TLD is a country code, such as .cz (Czech Republic).

2 To use e-mail a **client** computer needs an e-mail program to connect to a network **server**. The program can be stand-alone, e.g. Outlook Express, or Web-based, e.g. Yahoo. Stand-alone programs let you compose e-mail offline, but with Web-based programs you must be online.

3 E-mail uses two kinds of mail servers: an SMTP server, which **deals with** outgoing e-mail and a POP3 server, which deals with incoming e-mail. SMTP **stands for** Simple Mail Transfer Protocol; POP stands for Post Office Protocol.

4 If you send an e-mail to a friend in the same domain as you, your SMTP server simply sends it to the POP3 server in your domain, which adds it to your friend's inbox. If your friend has a different domain name, your SMTP server finds your friend's SMTP server using a Domain Name Service (DNS). When your friend's SMTP server receives the e-mail, it sends your e-mail to its POP3 server, which adds it to your friend's inbox.

---

3 Decide if the sentences are true (T) or false (F).

1 You cannot choose your own user name.   T/F
2 The domain name shows the network.   T/F
3 *.edu* and *.com* are TLDs.   T/F
4 A DNS sends an e-mail to a POP3 server.   T/F

4 Label the parts of the e-mail address.

1 _____   2 _____

Katrina123@hotmail.com

3 _____   4 _____

5 Look at these phrases from paragraph 4. What do the words in *italics* refer to?

1 … your SMTP server simply sends *it* to the POP3 server …
- a your e-mail
- b your domain

2 … *which* adds *it* …
- a the POP3 server/ your e-mail
- b your SMTP server/your domain

3 … *it* sends your e-mail …
- a the DNS
- b your friend's SMTP server

4 … to *its* POP3 server, …
- a your SMTP server
- b your friend's SMTP server

Vocabulary

6 Match the **highlighted** words in the text with the definitions (1–4).

1 means _____
2 a computer on a network _____
3 takes care of in some way _____
4 a computer that runs a network _____

7 Find the words in the text with the opposite meanings to these words.

1 offline _____   4 receive _____
2 Web-based _____   5 different _____
3 incoming _____

Speaking

8 Work in pairs. Look at the following European country codes. Discuss which countries they could stand for.

.at ■ .be ■ .bg ■ .de ■ .dk ■ .es ■ .fr ■ .gr ■ .hu ■ .it ■ .lu ■ .nl ■ .pt ■ .ro ■ .si ■ .uk

▶ *Get real*

Use an Internet search engine to find a list of Internet Country Codes. Pick any five countries that you do not know and find out where they are. Chose **one** country and find some information about it using your search engine. Report back to the class.

14 Sending files over the Internet

Before you start

1 Work in pairs and discuss the questions.

1 Have you used e-mail? Which program have you used?
2 Have you ever sent an attachment? What have you sent?

Reading

2 Read this information on attaching files in two different e-mail programs. <u>Underline</u> each action the user makes to send one attachment in each one. Write the number.

E-post Express   ☐
Mega Mail   ☐

● E-POST EXPRESS

You can attach a file while you are online or offline. Open the program and click Create a new mail message to go to your compose window. Click on the paperclip icon with the word Attach below it. An Insert Attachment dialog box appears, which shows your computer directory. Click on the file you want to send and then click the Attach button. The file and an icon appear in the Attach field. Send multiple files by repeating the procedure. The files can be any size but some servers will not accept files of more than one megabyte. To remove a file, click on the attachment with the right mouse button then click Re<u>m</u>ove. When you are finished, click Send.

● MEGA MAIL

Connect to the Internet to open your program and go to the compose screen. Click on Attach Files. A screen opens showing three Browse buttons. You can only send three attachments up to three megabytes in total. Click on the first Browse button. A Choose File dialog box appears, which shows your computer directory. Click on the file you want to send. The file name appears in the File <u>n</u>ame drop-down list box. Click Open. The Choose File dialog box disappears and the file appears in the file field of the Attach Files screen. Click Attach files. A screen appears telling you that the file is being attached and then another screen appears when the program has attached the file. To add more files, click Attach More Files and the Attach Files screen will reappear. When you have finished, click Done. Your compose screen reappears, listing the name of the attached file with an icon next to it. Click Send.

3 Which information (1–6) is the same for E-post Express and Mega Mail, and which is different? Write S (same) or D (different).

1 You can send up to three megabytes of data. ☐
2 A dialog box appears, showing the computer directory. ☐
3 You can attach and send up to three files. ☐
4 You have to be online to attach files. ☐
5 Click Send when you want to send your e-mail. ☐
6 The program shows an icon next to the attached file. ☐

Vocabulary

4 Find the words in the text that mean:

1 a series of steps (E-post Express) _____
2 take off or take away (E-post Express) _____
3 something you click to start an action (E-post Express) _____
4 goes away suddenly (Mega Mail) _____
5 come back into view (Mega Mail) _____
6 finished (Mega Mail) _____

5 Match the prefixes (1–4) with their meanings in the box. Two prefixes have the same meaning.

again ■ not/negative ■ before

1 *re-* means _____
2 *pre-* means _____
3 *dis-* means _____
4 *un-* means _____

6 The words below can take the prefixes *dis-*, *pre-*, *re-* or *un-*. Write a prefix in front of each word. (Some can take more than one prefix.)

1 _*re*__write   6 _____titled
2 _____do   7 _____appear
3 _____like   8 _____format
4 _____agree   9 _____finished
5 _____view   10 _____start

Writing

7 Write a description of another method of sending a document or a picture to someone, e.g. by post, by fax, by picture messaging on a mobile phone. Use the texts in Exercise 2 to help you.

▶ ### Get real
Attach three files from your computer to an e-mail. Report back to the class on how long it took you to attach and send the files, and whether anything went wrong.

Before you start

1 Work in pairs and answer the questions.

1 What is an e-card? When do you think you send an e-card to someone?

2 What can you download from the Internet? Make a list.

Reading

2 Circle the answers *yes* or *no*. Read the text to check your answers.

1 Can you send music in an e-card?	yes / no
2 Do you have to save an e-card to view it?	yes / no
3 Do you pay for freeware programs?	yes / no
4 Can you download a movie from the Internet?	yes / no

▢ Guide to downloading files

Viewing websites
You can view many interesting websites with your browser. Some let you view and send e-cards for birthdays, holidays or other special occasions using your e-mail program. An e-card can contain pictures, cartoon animations, or play songs. You can type your own personal message on the card, change the music, preview it, or send it as a screen saver. Most e-cards open automatically in your e-mail, others give you a link to click. You usually view e-cards like a standard Web page.

Downloading programs
You can download computer programs, games and utilities, such as virus protection programs. Some of these programs are shareware, which means you pay a fee if you keep the program, or freeware, which have no fee. To download a program, you save it on your computer. After you click the download button, the Save As dialog box appears. Choose the location where you want to save the file and click Save. It can take anything from a few seconds to a few hours for a download to complete.

Downloading e-mail attachments
You can view e-mail attachments on the Internet or you can save them onto your computer. To open an attachment your computer needs a program that can open it. If your computer does not have compatible software, you cannot open the attachment. All digital files have a file extension that shows you the file format, for example *.avi* for video, *.doc* for MS Word files and *.mpeg* for music files.

3 Match the first part of the sentence (1–6) with the second part (a–f).

1 To view your e-card,	a in MS Word.
2 You cannot keep shareware	b online or offline.
3 To download a computer program	c lets you save an attachment.
4 You can view an attachment	d if you don't pay for it.
5 You cannot view a video program	e open your e-mail program.
6 The Save As dialog box	f you click Save As.

Vocabulary

4 Which of the words and phrases in the box are specific to IT and which are used in general English? Use the Glossary or a dictionary to help you.

> animations ■ automatically ■ compatible ■ complete ■ download ■ file extension ■ standard ■ utilities

Speaking

5 Work in groups. Match the messages with the occasion and the person.

Occasion Hallowe'en, arranging a meeting, apology, thanking someone, missing someone, birthday
Person granddad, friend, aunt, brother/sister, neighbour, girl/boyfriend

❶ Have a great **78th**. Careful when you blow out the candles!

❷ Hi! How are things? See you in town at 3.00. Be good!

❸ Thank you for the socks. They'll be useful this winter. Your loving nephew.

❹ Scary card, eh? Good idea for the party?

❺ Love you & really want to see you soon! XXX

❻ I am so sorry for breaking your window. Please accept my deepest apologies.

Now make more messages for different occasions and people.

▶ ## Get real
Either review two or three e-card websites or go to a shop with a good selection of cards. Which occasions are there cards and e-cards for? What do the cards and e-cards offer (e.g. pictures and sounds)? Were the websites easy to use? Report back to the class and discuss what you found.

16 Music on the Internet

Before you start

1 Tick (✓) the kind of music you like. Make a list of other types of music.

pop ☐   rock ☐   classical ☐   jazz ☐

2 What is good or bad about downloading music from the Internet? Make a list.

Reading

3 Milos (M) is a music fan. Kamila (K) works in the music industry. They are in a chat room. Read the dialogue and tick (✓) the topics they talk about.

1 Making copies of songs from the Internet. ☐
2 How much money the music industry loses. ☐
3 How Napster sent music to people. ☐
4 What peer-to-peer music sharing is. ☐
5 How to stop peer-to-peer sharing. ☐
6 Which are the best legal music websites. ☐

M Downloading music is great. I can get all the songs I like, when I want to get them.

K That's true, but if you don't pay for it, you're breaking copyright law.

M Really? Why is it against the law?

K Well, getting music for free costs the music industry billions in lost income, so we have less to spend on new bands and singers.

M Was that the problem with Napster?

K Yes. Napster created a file-sharing system using the MP3 audio format. People could connect to a central location and others could then download their files using the server. Napster closed down in 2001 because it was breaking the law.

M I see, but peer-to-peer music swapping is legal, isn't it? It's just two people sharing music – they're not using a central server.

K No, it's still illegal, I'm afraid.

M Actually, peer-to-peer isn't that great – you don't get much choice because it depends on who's online at the same time as you. Can you download music legally?

K Yes, there are several Web-based music services that charge a fee. It's really worth paying. The choice and quality of the music is better and they offer other services such as music reviews and chat rooms. Try one!

4 Label the diagrams *central location* and *peer-to-peer*.

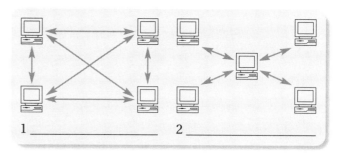

1 _____   2 _____

5 Match the first part of the sentence (1–6) with the second part (a–f).

1 The record industry loses money
2 Napster used
3 You have to pay money
4 The best music websites
5 Peer-to-peer sharing

a to join a legal music website.
b give a lot of music services.
c is illegal.
d the MP3 audio format.
e because of peer-to-peer sharing.

Vocabulary

6 Find the words in the text that mean:

1 money people receive for work _____
2 related to sound _____
3 someone of the same type/group _____
4 exchanging something with someone _____
5 dividing something between people _____
6 against the law _____
7 money you pay for a service _____

Speaking

7 Work in groups. What do you think about copyright laws and downloading music from the Internet?

Writing

8 Write two paragraphs on downloading music for free. The first should give the record companies' and artists' views and the second should give music listeners' views.

▶ ### Get real
Visit one or more websites offering music downloads for a fee. Make notes on the type of music they give you, the services they offer, and the cost. Report back to the class, saying which site you think is the best, and why.

Before you start

1 Label the shapes and lines with the words in the box.

a _____ **b** _____

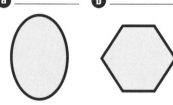

c _____ **d** _____

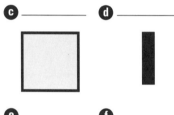

e _____ **f** _____

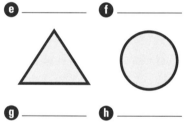

g _____ **h** _____

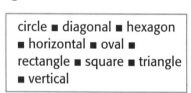

i _____

> circle ■ diagonal ■ hexagon ■ horizontal ■ oval ■ rectangle ■ square ■ triangle ■ vertical

2 Write the words from Exercise 1 under the correct heading.

Noun

_____
_____
_____
_____

Adjective

_____
_____
_____

Noun and adjective

_____
_____

Reading

3 A computer virus has damaged this text and put the paragraphs in the wrong order! The headings (1–6) are in the correct order. Use them to number the paragraphs.

1 DTP programs and what they do
2 Templates and Web pages
3 Making changes to text
4 DTP programs and word processors
5 Using graphics
6 Moving text and graphics on a page

☐ DTP programs let you work with graphics: you can draw shapes, fill them with text or colour, insert graphics or special characters from the program, or import them from another program, and you can move them all easily around the page.

☐ While DTP programs and word-processing programs have a lot of similar commands and tools, DTP programs have one important advantage: what you see on the screen is exactly what you get when you print your document.

☐ There are many things you can do with text and graphics: you can use an align command to put them in a straight line, horizontally or vertically, and a rotate tool lets you turn them around. You can bring text to the front of a shape or graphic or send it to the back so that you can't see it. You can also wrap text around a picture or inside a shape, like in this reading.

☐ Desktop Publishing (DTP) programs, such as Adobe PageMaker and QuarkXpress, let you combine text and graphics in creative ways to produce stylish greeting cards, holiday brochures, business cards, newsletters, sales catalogues, calendars and many other documents.

☐ The tools and commands in DTP programs give you a great deal of control over text. For example, you can make word and character changes, such as changing the space between words in a text without changing the font size, or changing the space between characters to make them look neater. These choices are useful when you only have a small space to work in.

☐ These programs also let you make a template of your document so you do not have to remake the whole document each time you want to change the text or the pictures. Many DTP programs let you change the file format of your design into a Web page, too.

4 Read the text in the correct order. Decide if the sentences are true (T) or false (F).

1 You cannot type letters in a DTP program. T/F
2 You can use a template to save time. T/F
3 DTP programs print exactly what you see on the screen. T/F
4 It is difficult to control the text in a DTP program. T/F
5 It is impossible to change the spaces between words. T/F
6 A rotate tool lets you turn text around. T/F

5 Complete the chart with the correct command or description of each action from the box.

> align horizontally ■ increase space between words ■ align vertically ■ bring to front ■ rotate ■ fill ■ text wrap ■ decrease space between characters

	Before	After	Command
1	▬	╱	
2	▬ ▬ ▮	▬ ▮ ▮	
3	▬ ▬ ▬	▬ ▬ ▬	
4	▽	▽	
5	H⬭rld!	Hello World!	
6	XOXOXOXOXOXOXO XOXOXOXOXOXOXO XOXOXOXOXOXOXO XOXOXOXOXOXOXO	XOXOXOXOXOXOXO XOX ⬭ XOXO XOX XOXO XOXOXOXOXOXOXO	
7	H e l l o	Hello	
8	IT is fun!	IT is fun!	

Vocabulary

6 Complete the puzzle with words from the text.

1 looking smart and new
2 letters and words
3 a document pattern or plan
4 join together
5 bring from another program
6 get from the program you are using
7 a drawing or plan of something new

7 Match the document types (1–7) with the descriptions (a–g).

1 greeting card   **a** information for tourists showing hotels, resorts, etc.

2 holiday brochure   **b** a table that shows days and months

3 business card   **c** a small card with a person's name and a company name

4 newsletter   **d** a card for a special occasion

5 sales catalogue   **e** a page on the World Wide Web

6 Web page   **f** a small newspaper for a specific group of people

7 calendar   **g** a book of advertisements from a commercial company

Speaking

8 Work in pairs. Describe the picture.

9 Now design your own picture with words and text. Describe it to your partner but do not let him/her see the picture. Your partner has to draw the picture. You could then try designing your picture and text in a DTP program!

> ▶ **Get real**
> Bring in some holiday brochures, newsletters, business cards, sales catalogues or any other printed documents. Look the design of the documents. Discuss how you could change and improve them using the commands and tools of a DTP program.

Before you start

1 Work in groups and discuss the questions.

1 Do you like paintings, photographs or computer art best? Why?

2 'A picture paints a thousand words'. Do you think this saying is true?

Reading

2 Read the text. Write the words in the box under the correct heading.

> clip art ■ JPEG ■ Web pages ■ TIFF ■ adjust ■
> special effects ■ PICT ■ cut ■ EPS ■
> digital cameras ■ paint ■ scanners ■
> create new ■ erase ■ GIF ■ paste

Image editing	Image formatting	Image sources
_____	_____	_____
_____	_____	_____
_____	_____	_____
_____	_____	_____
_____	_____	_____
_____		

People who use DTP programs often have an image-editing program, such as Adobe Photoshop or CorelDRAW, on their desktop. You can get images from many sources : you can draw or paint your own new images, import clip art and other images from CD-ROMs, and save images or pictures from Web pages. You can also transfer photographs from a digital camera or use images scanned into your computer from a scanner.

All image-editing programs have similar tools and commands. You can do much more with your pictures and images than you can with a DTP program. You can erase parts of an image or cut and paste them onto another image, adjust the brightness, paint patterns or lines and add all kinds of special effects .

You can save an image in many different file formats. GIF, for example, is used for animation and is a popular choice for Web pictures, but has fewer colours than other formats. JPEG is good for photographs and downloads quickly from the Web, but it can lose image data when you save it. Apple Macintosh designed PICT for the MacOS, but TIFF is a good cross-platform format that you can use with many operating systems. If you can't use an image in a DTP program because the program doesn't let you, you can often export it in EPS format from your imaging program, without losing any picture quality.

3 Write the best file format to use for saving the image (not TIFF).

1 You have scanned in a photograph into your computer. _____

2 You want to export your picture to another document. _____

3 You want to make a cartoon for the WWW. _____

4 Your computer runs the MacOS. _____

4 Look at these questions about imaging software. Circle *yes* or *no*.

1 Can I copy images from Web pages?   yes / no
2 Can I put two images together?   yes / no
3 Can I make pictures darker?   yes / no
4 Do JPEGs ever lose picture quality?   yes / no
5 Is PICT an operating system?   yes / no

Vocabulary

5 Complete the sentences (1–6) with the highlighted words and phrases in the text.

1 Making a picture look softer is just one of many _____ in Photoshop.

2 It's usually quicker to use a _____ image than to draw it yourself.

3 Scanners are _____ peripherals that you can use with any operating system.

4 You can get images for your website from many different _____.

5 The picture was too bright so I had to _____ the brightness.

6 You can remove that ugly building in the picture with the _____ tool.

Speaking

6 Work in pairs. You are going to make an image to put on your school Web page for the following events/things. What kind of image would you make for each one?

- a school sports day
- an end-of-school party
- a school trip to the zoo
- an advertisement for the Computer Club
- a 'No Smoking' sign
- an advertisement for the English Club

> ▶ **Get real**
> Look at some of the pictures and images in magazines and some created by computer programs (type 'computer generated art' in your search engine). Which colours go well together? Report back to class on your favourite images and websites.

Before you start

1 What things don't you like or annoy you about websites? Make a list.

Reading

2 Look at Matej's *Top 10 Web page annoyances* on his home page. Which ones are similar to the annoyances you talked about in Exercise 1?

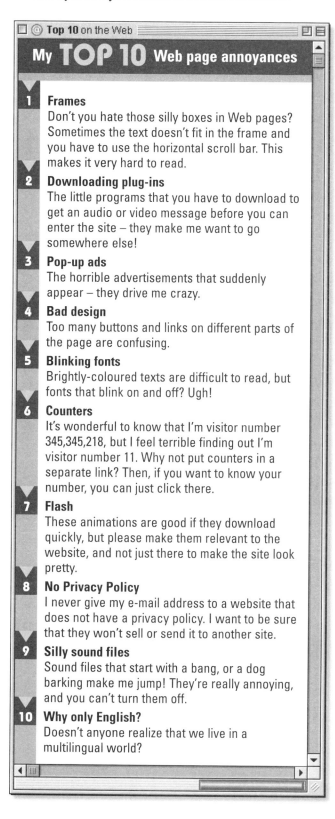

□ @ Top 10 on the Web

My TOP 10 Web page annoyances

1 Frames
Don't you hate those silly boxes in Web pages? Sometimes the text doesn't fit in the frame and you have to use the horizontal scroll bar. This makes it very hard to read.

2 Downloading plug-ins
The little programs that you have to download to get an audio or video message before you can enter the site – they make me want to go somewhere else!

3 Pop-up ads
The horrible advertisements that suddenly appear – they drive me crazy.

4 Bad design
Too many buttons and links on different parts of the page are confusing.

5 Blinking fonts
Brightly-coloured texts are difficult to read, but fonts that blink on and off? Ugh!

6 Counters
It's wonderful to know that I'm visitor number 345,345,218, but I feel terrible finding out I'm visitor number 11. Why not put counters in a separate link? Then, if you want to know your number, you can just click there.

7 Flash
These animations are good if they download quickly, but please make them relevant to the website, and not just there to make the site look pretty.

8 No Privacy Policy
I never give my e-mail address to a website that does not have a privacy policy. I want to be sure that they won't sell or send it to another site.

9 Silly sound files
Sound files that start with a bang, or a dog barking make me jump! They're really annoying, and you can't turn them off.

10 Why only English?
Doesn't anyone realize that we live in a multilingual world?

3 Write the number of the annoyance that matches each sentence.

a Websites should have rules about giving e-mail addresses to other sites. □

b Web pages should have options for different languages. □

c Animations should be about the same subject as the Web page. □

d Web pages should not need special programs to run. □

e The text should not blink on and off. □

f Dividing a Web page into a lot of small frames is bad design. □

Vocabulary

4 Find the opposites of the adjectives (1–6) in the text.

1 clear _____
2 very bad _____
3 unrelated _____
4 ugly _____
5 uncertain _____
6 monolingual _____

5 Complete the sentences (1–5) with the words in the box.

confusing ■ else ■ fits ■ relevant ■ scroll bar

1 This is a well-designed Web page. Everything _____ well on the screen.

2 Web surfers will go somewhere _____ if the page doesn't appear quickly.

3 That animation is good because it is pretty and it is _____ to the Web page.

4 That website is very _____ because I couldn't find the Back button.

5 It's annoying when you have to use the horizontal _____ to see all of the text.

Speaking

6 Work in pairs. Do you agree or disagree with Matej's list? Rank your top five annoyances from the text: 1 = most annoying, 5 = least annoying.

▶ Get real

Work in groups. Look at some websites for business, education, entertainment, or information. Make notes on the differences in design between them. Look at how they organize navigation bars, the categories they use, and how many pictures and animations they use. Report back to the class and make a class list of design features for each category.

Before you start

1 Work in pairs and discuss the questions.

1 Why do people have personal websites?

2 Have you ever visited anyone's personal home page? What was it like?

Reading

2 What do you know about Web page design? Answer our Internet Quiz then check your answers by reading the article.

3 Read the paragraph *Beginning HTML* again. Match the first part of the sentence (1–6) with the second part (a–f).

1 HTML tags tell the browser where

2 A Web-authoring program

3 FTP is a set of transfer rules

4 Web page designers use

5 See your Web page on the Internet

6 Tags are

a that are used to upload Web pages to a server.

b the text, graphics and animations go.

c by uploading it to a Web server

d HTML tags to create Web pages.

e a set of codes in HTML.

f writes HTML tags for you.

4 Look at the *Home Page Hints* again. Write the hint numbers in the correct column.

Do ...	Don't ...
____	____
____	____
____	____
____	____
____	____
____	____

Internet Quiz

1	You don't need to learn HTML to make a Web page.	T / F
2	Making a Web page is like designing a printed page.	T / F
3	You shouldn't use a lot of different colours in your texts.	T / F
4	Having a lot of pictures and animations on Web pages is great.	T / F
5	Surfers like reading on the Net.	T / F

Web page design

Many of our readers are setting up home in cyberspace. Read how you can, too!

Beginning HTML

Web designers use Hyper Text Markup Language (HTML) to create and format Web pages. HTML uses a set of codes, called tags, to **structure** a Web document that will run in a browser. There are hundreds of tags you can use to format text, insert graphics, animations, sound and video. But you do not need to understand HTML to make your own personal home page. Many word-processing, desktop publishing and **Web-authoring** programs will **generate** HTML tags for you. To upload, or copy, your Web page to a Web server, use the server's File Transfer Protocol (FTP).

Home Page Hints

It's your 'cyberhome', but remember that websites are different from books or magazines. Think about these suggestions to make people want to stay.

1 Use a **navigation bar** to organise your hyperlinks to other pages.

2 **Hyperlinks** also let visitors navigate up or down long pages.

3 Keep your use of colour and buttons **consistent**. If a Next Page button is a pink circle, all Next Page buttons should be the same, and in the same place on the screen.

4 If you use a lot of animations, your Web page will take a long time to download.

5 If you use a lot of graphics, animations and text your Web page will be too busy.

6 It's difficult to read a text that's next to an animation.

7 Keep texts short and simple! Surfers don't like reading on a computer screen much.

8 It's not easy to read multi-coloured text.

9 Lots of bright colours look nice at first, but often give people headaches!

10 Make sure you use a spell check and use good grammar.

11 Try not to use too much slang. People who visit your site may not understand.

12 Don't be afraid to be **original**. Good websites have something that is different about them and that comes from you!

Vocabulary

5 Match the `highlighted` words and phrases in the text with the definitions (1–7).

1 connections to a Web page or part of a Web page _____
2 make or produce _____
3 plan or build a Web page _____
4 a group of organised Web links, usually in a line _____
5 does not change, always the same _____
6 the type of software that helps create Web pages _____
7 new, not done before _____

6 Complete the sentences (1–8) with the words in the box.

> busy ■ consistent ■ generate ■ home pages
> ■ structure ■ surfers ■ upload
> ■ Web-authoring

1 That Web page is much too _____. I don't know what to look at.
2 An FTP server is a computer that lets you _____ files to the Internet.
3 The buttons on this page are not _____ with the button on the last page.
4 Net _____ never like reading a lot of text on the screen.
5 _____ software means you don't have to learn HTML to make a Web page.
6 Many students have their own _____ on the World Wide Web.
7 HTML creates the _____ for Web pages to run on a browser.
8 Web-authoring programs _____ HTML tags for you.

7 Tick (✓) the sentences that use informal English.

1 Get real, people. Frames are a big no-no. ☐
2 Designing a Web page needs careful planning. ☐
3 I think sound files are cool. ☐
4 Multi-coloured blinking fonts look terrible. ☐
5 That Web page is mega ugly. ☐
6 The text has too many grammatical errors. ☐

Speaking

8 Work in groups. How would you design your personal website? What graphics, images and colours would you use? What would you say in your text? How many pages would you have? What would you call the links on your navigation bar to show the different pages? Use the sample pages in Exercise 9 to help you.

Writing

9 Look at the two home pages below. Write two paragraphs, describing what is good and bad about each home page.

Freddy's home page is fun but badly designed because it has/uses …
It is confusing because it is …/there are …
Jana's home page is well designed because it has …
It is easy to navigate because it is …/there are …

▼ Freddy's Home Page

▼ Jana's Home Page

▶ Get real

Review the texts in Unit 19 and in this unit. Visit a website or home page of your choice. Make notes on what is good and bad about it. Report back to the class and make a class file of good and badly designed sites for people to visit.

Before you start

1 Work in groups and discuss the questions.

1 How are books and CD-ROMs different?
2 Have you ever used CD-ROMs to help you study? Do you prefer them to books?

Reading

2 Match the parts of the CD-ROM with the information they provide.

a The history of multimedia ☐
b Education and entertainment ☐
c What is multimedia? ☐
d Business and industry ☐

1 Multimedia is any computer application that integrates text, graphics, animation, video, audio or other methods of communication. Multimedia is different from television, books or cassettes because it lets you interact with the application. You can click on a word to make a picture appear, or click on a picture to start a video.

2 Multimedia became more popular after the mid-1990s when the price of hardware began to fall. Then people started using it in industry, business, education, entertainment and for other purposes. Today, we can find multimedia at home, in school, at work, in public places, such as libraries, and on the Internet.

3 In business, advertisers use virtual reality in multimedia applications to advertise their products in three dimensions (3-D). Using multimedia for graphs and tables is now the best way for managers to present company results. In industry, pilots learn to fly using multimedia simulations of real situations, and scientists simulate experiments with dangerous chemicals in safety. Publishers are also producing interactive magazines, called e-zines, and e-books online.

4 In education, students study interactive CD-ROMs at their own speed and explore topics creatively by clicking on related links. Teenagers have played computer games for years, but many multimedia applications combine education and entertainment and they let them visit virtual worlds or change the ending of films.

3 Complete the sentences with one way these people use multimedia applications.

1 Advertisers _____
2 Managers _____
3 Pilots _____
4 Scientists _____
5 Publishers _____
6 Students _____
7 Teenagers _____

4 Match the first part of the sentence (1–5) with the second part (a–e).

1 People like using multimedia
2 Multimedia combines
3 Most educational CD-ROMs
4 Prices of multimedia hardware
5 Students like learning about new topics

a started falling around 1995.
b using interactive multimedia.
c many different ways of learning.
d integrate audio, video and text.
e because it is interactive.

Vocabulary

5 Complete the sentences (1–6) with the words in the box.

application ■ integrated ■ related ■ simulation
■ 3-D ■ virtual

1 The image is in _____ so you can look at it from all sides.
2 All the links in this e-zine are _____ to football.
3 This _____ makes you think you really are flying to the moon.
4 Some shopping websites use _____ reality to advertise their products online.
5 The application is _____ because it combines many features.
6 Any program that carries out a specific task for a user is an _____.

Speaking

6 Work in pairs. Describe your favourite CD-ROM (or other method of studying). What can you learn from it? Describe how it integrates text, images, and other features such as animations, video, audio and Internet links.

> ### Get real
> Find an e-zine on the WWW on a topic you are interested in. Note how it is different to reading a paper magazine. Report back to the class.

22 E-commerce

Before you start

1 Work in pairs and discuss the questions.

1 Have you ever heard of e-shopping? What do you know about it?

2 Do you know anyone who has bought anything online? What did they buy?

3 What do you think are the advantages and disadvantages of e-shopping? Make a list.

Reading

2 Read part of an interview about e-commerce. Match the questions (a–e) with the correct paragraphs (1–5).

a How does e-commerce work? ☐

b What's the future for e-commerce? ☐

c Do customers like shopping online? ☐

d What kind of business do you run? ☐

e What do e-tail stores need to succeed? ☐

£¥$€£¥$€ Business news and views £¥$€£¥$€

1 We sell mobile phones and accessories, and we only operate online. We're a B2C business. That means 'business to customer', so we don't sell to other businesses – that's B2B. We're obviously not C2C either, which is individual people selling to each other online.

2 Yes, it's becoming very popular and successful. It's world-wide shopping, 24 hours a day, 365 days a year. It's so convenient – people can browse through online catalogues, compare prices easily, and there's less paperwork, so it's cheaper for the retailer. We can pass these savings onto the consumer.

3 Well, the best sites, or e-tail stores, have an electronic storefront giving categories that are easy to understand. You can read reviews about the products, go to chat rooms to talk about them, and when you've made your choice, simply click your mouse and add it to your electronic shopping cart.

4 The retailer needs to build consumer confidence. You need a website that is easy to navigate and it must download quickly. You need customer support services, things like FAQs (frequently asked questions), information about the order, and guarantees about delivery. A secure server for transactions using credit cards and a privacy policy are also very important.

5 I think everyone will shop online soon. All e-tail stores will use virtual reality to sell their goods – it's going to generate billions of euros.

---

3 Tick (✓) the features of the best e-tail stores.

1 have slow downloads ☐

2 have an electronic shop window ☐

3 have somewhere to put to your goods ☐

4 inform the customer about the order ☐

5 need a lot of paperwork ☐

6 have a place for people to talk ☐

7 give product reviews ☐

8 use a safe Web server for payment ☐

9 cannot say when goods will arrive ☐

10 let people ask questions ☐

Vocabulary

4 Find the words in the text that mean:

1 extra, additional products (paragraph 1) _____

2 work, do business (1) _____

3 matches someone's needs (2) _____

4 someone who sells (2) _____

5 customer (2) _____

6 help (4) _____

7 promises (4) _____

8 buying and selling (4) _____

5 Complete the table with the adjective or noun in the text (and questions).

Noun	Adjective
_____	commercial
convenience	_____
success	_____
_____	confident
security	_____
_____	private

Writing

6 Work in pairs. What questions do you think an online consumer will have? Think of five FAQs for an online business, e.g. about the type of business, the products offered, payment, security.

• Write the questions.

• Give them to another pair.

• Imagine you run an online business and answer the questions you receive.

• Write the answers under the questions and give them back to their authors.

▶ ### Get real

Look in your high street and find big or small stores that have an e-tail site. Choose one or two and look at their websites. What kind of services do they give their Internet customers? Compare your findings with other students. Which are the best e-tail stores?

23 Chat rooms

Before you start

1 Work in groups. Make a list of:
- five topics you can talk about when you first meet people
- five topics you shouldn't talk about when you first meet people.

Reading

2 Read the opinions about chat rooms. Which ones do you agree with?

We got a lot of letters in response to our article last week on Internet chat rooms. Here is a random selection.

1 Why do people like them? They're boring! It's just a group of people talking nonsense! My son doesn't go out or meet real people any more!

2 Some make you register for a free trial membership, so you have to send your real name and e-mail address. You have to read the agreement carefully – sometimes if you don't cancel before your trial ends, they send you a bill! I think this is unfair.

3 You should warn parents about them. People use nicknames – they call them 'handles' – so you don't know who they are. Tell teenagers never to give out personal information, especially their name, home or school address or telephone number – and they must never agree to meet anyone from a chat room. It could be dangerous.

4 I think they're good for practising English in real time – that's when everyone is online and 'talking' at the same time as you. I like expressing my feelings with those cute emoticons, too. If you only talk about your family, the weather, sport, school subjects and other small talk topics, I think they are amusing and harmless.

5 Most only have text boxes for messages, but chat rooms that support voice and video chat are the best, if you have the right hardware and software that is!

6 If people can't find a chat room they like, they can create one of their own. I set up my own online community. I think this is fantastic and more people should do the same.

3 Decide if the sentences are true (T) or false (F).
1 You can set up your own chat room.   T/F
2 You must be online to go into a chat room. T/F
3 Anyone can use a video chat room.   T/F
4 You can say how you feel with a symbol.   T/F
5 You have to pay for some chat rooms.   T/F

4 Tick (✓) the things you can say in a chat room. Put a cross (✗) for the things you shouldn't say.
1 My surname is Sukova. ☐
2 Which school do you go to? ☐
3 My e-mail address is SK45@yahoo.com. ☐
4 Do you like hip hop music? ☐
5 Is it raining where you are? ☐
6 Have you ever been to England? ☐
7 My telephone number? Sure, it's 234 6358. ☐
8 My other handle is 'bluebird'. ☐
9 What's your favourite subject? ☐
10 OK. Let's meet at the shopping mall at 3.00. ☐

Vocabulary

5 Read the text and make six two-word phrases. Then write the meanings in your language. Use the Glossary or a dictionary to help you.

community ■ information ■ membership ■ room ■ talk ■ time

1 personal _____   _____
2 trial _____   _____
3 chat _____   _____
4 small _____   _____
5 real _____   _____
6 online _____   _____

Speaking

6 Work in pairs. Match the emoticons with the meanings and descriptions in the box. Do you know any others?

1 :(2 :((3 :D   4 :)
5 >:(6 B)   7 :| 8 :O

shouting ■ I don't care! ■ sad ■ very angry ■ cool sunglasses smiley ■ laughing ■ very sad ■ happy

▶ ## Get real
Ask the people in your family what they talk about when they first meet somebody. Take each topic and think of questions in English that you can ask about them. Make a class list of small talk topics and questions.

Before you start

1 Look at the definition of *etiquette*. What do you think *Netiquette* is?

> **etiquette** /'etɪket, -kət/ *n* [U] formal rules of correct and polite behaviour in society or among members of a profession

Reading

2 Read the Web page about Netiquette and check your answer to Exercise 1. Then write the headings (a–d) above the correct paragraphs (1–4).

a Rules for talking online ☐
b Invading privacy ☐
c The Golden Rule ☐
d Culture and Netiquette ☐

☐ Online Netiquette   回目

1 _____

People in the West usually shake hands when they first meet. Good friends in Middle Eastern cultures kiss each other three times on the cheeks. The Japanese bow their heads to show respect and the Thais bring their hands together in front of their face. The online community, too, has its own culture and **customs**. Good Internet behaviour is called Netiquette.

2 _____

The Internet is an international **means** of communication where you can talk to people online. Asking questions is fun but making jokes about people from other cultures can lead to misunderstanding and bad feelings. Sending hurtful or insulting messages, or flames, to people is bad behaviour. Bad language is not cool. Everyone is happy when people are friendly.

3 _____

Netiquette includes more than good spelling and grammar. Typing in all upper case is bad as it is the same as SHOUTING. Not starting your sentences with capital letters is lazy. Sending e-mails with 'Hello' and 'Thank you' is nice. The Golden Rule is 'Treat others in the same way that you like to be treated.' Remember, real people read what you type!

4 _____

It is also bad Netiquette to send people spam. This kind of **unsolicited** e-mail means people have to cancel something that they did not ask for in the first place. When you use Cc: **instead of** Bcc: you send other people's e-mail addresses without their permission. This is an **invasion** of their privacy and breaks the Golden Rule.

---

3 Make questions from the text for these answers.

1 _____
   When they meet someone for the first time.

2 _____
   To show respect.

3 _____
   Netiquette.

4 _____
   Real people.

5 _____
   It sends an e-mail address when you haven't asked the owner.

4 Tick (✓) the things which are good netiquette and put a cross (✗) by those that are bad netiquette.

1 correct spelling ☐
2 using Bcc: instead of Cc: ☐
3 sending e-mail that people do not want ☐
4 greeting someone in an e-mail ☐
5 making jokes about people's culture ☐
6 typing in capital letters ☐
7 flaming people ☐

Vocabulary

5 Complete the sentences (1–5) with the highlighted words in the text.

1 Telephones and postal services are both _____ of communication.
2 Different cultures usually have very different _____.
3 Please don't copy my e-mail to other people. It's an _____ of my privacy.
4 Try asking interesting questions _____ trying to think of funny things to say.
5 People on the Internet are always complaining about _____ e-mail.

Speaking

6 Work in groups. Discuss the 'rules' of etiquette in your country. Think about things like greeting, saying goodbye, queueing, visiting someone's home.

Writing

7 Write a paragraph describing common customs and behaviour in your culture.

> ▶ ***Get real***
>
> Find out some interesting cultural customs of one of the following: the USA, Britain, Japan, the Middle East, Thailand. Make notes on what you found and report back to the class.

Before you start

1 Work in pairs and discuss the questions.

1 Have you ever seen a chart like the one in Exercise 2?
2 Where have you seen one?
3 What kind of information did it give you?

Reading

2 Look at the algorithmic flow chart below. Answer the questions.

1 Which computer commands does it show? _____ and _____
2 How many decisions does the computer make? _____
3 After the user clicks Save, how many times does the user input data? _____

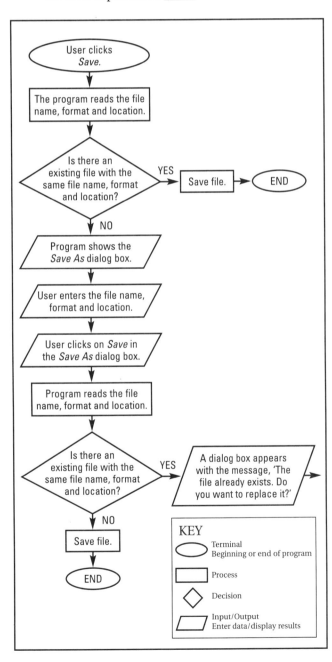

3 Read the text about computer programming. Write the number of the paragraph that gives you the information.

a a description of machine language ☐
b the greatest problem for computer programmers ☐
c the names of three high-level computer languages ☐
d a description of an algorithm ☐
e different uses of computers in our lives ☐

1 The diagram on the right shows part of a simple algorithmic flow chart for the Save command in a computer program. An algorithm is a set of logical rules that we use to solve a problem. Computer programmers often use algorithms to plan their programs, but the only language a computer understands without translation is machine language. This uses the binary system of 1 and 0, which matches the electrical positions 'on' and 'off'. We can also show these numbers in English by *Yes/No* or *True/False*.

2 Machine language is a low-level language and is very difficult to write. Over the years, computer scientists have developed many high-level languages, such as BASIC, C++ and Java. These languages use a computer code that is similar to English, which makes them easier to learn. A computer program is just a set of coded instructions. A computer translates the code into machine language to complete a specific task. A computer receives input, processes data and produces results, or output, according to the program code.

3 We use computers in many parts of our lives, and not just in schools or for the Internet. There are computers in all kinds of electrical devices, from mobile phones to washing machines. We can find them in banks, supermarkets and cars. When programmers write programs, they have to plan carefully for every possible kind of error a computer user can input into the computer. It is planning for the random behaviour of humans that makes programming so much fun.

4 Circle the answers *yes* or *no*.

1 Programmers use algorithms when
writing programs.   yes/no
2 Programmers write programs using
the numbers 1 and 0.   yes/no
3 Machine language is a high-level
language.   yes/no
4 We only find computer programs in
computers.   yes/no

5 Answer the questions.

1 What is an algorithm?
2 What does a binary system consist of?
3 Why are high-level languages easy to learn?
4 What do computers do with code?
5 Why must programmers plan carefully?

Vocabulary

6 Complete the sentences (1–6) with the words in
the box.

according to ■ behaves ■ devices ■ errors
■ input ■ output

1 _____ the bank machine, I have no money
in my bank account.
2 Video and digital cameras are other _____
that use computers.
3 _____ is any result a computer displays on
a screen or prints from a printer.
4 A computer receives _____ from users
when they click on a command.
5 I made too many _____ in my test so I got
a bad grade.
6 Not everyone _____ logically when things
go wrong with a computer.

7 Choose the correct word.

1 Most people can easily learn a _____ language
and become programmers.
   a low-level   **b** high-level   **c** binary
2 You can use a mouse or keyboard to _____ data
into the computer.
   a output   **b** input   **c** process
3 Some children _____ very badly when they can't
have something they want.
   a process   **b** solve   **c** behave
4 You can draw _____ for many simple
procedures.
   a an algorithm   **b** a code   **c** data
5 Computers _____ programming language into
machine language.
   a use   **b** develop   **c** translate
6 A computer can _____ large amounts of data at
very high speeds.
   a process   **b** result   **c** complete

Speaking

8 Work in groups and discuss the questions.

1 What kind of errors do you make with
computers?
2 How do you behave when things go wrong with
a computer (or any machine)?
3 How do different people you know behave
when things go wrong?

Writing

9 Draw a flow chart like the one in Exercise 2.
Follow these steps:

• Choose a simple procedure from the ones in
the box below (or a similar one of your own).
• Break the procedure down into all the steps
that you have to follow. Think about where the
process starts and ends, and the input from you
and from the outside. When you make a
decision, think of when you say 'yes' and when
you say 'no', and what happens next.
• Write exactly what happens at each stage.
• Draw the flow chart, putting your text into the
different shapes in Exercise 2.
• Show your flow chart to another student. Does
he/she agree with your steps?

making a cup of tea or coffee ■ making a
telephone call ■ sending a text message ■
answering the door ■ playing a cassette or CD
■ putting on the washing machine

▶ *Get real*

Make a list of all the devices that use
computers in your home, in your school,
shops, libraries, banks and offices. During a
day, use your list to make a note of every time
you use something that contains a computer
program. Report back to the class with your
list.

Before you start

1 Make notes to answer the questions.

1 What is videoconferencing used for?
2 What are the advantages and disadvantages of videoconferencing?

Reading

2 Read the text quickly and match the headings (a–d) with the video screens (1–4).

a How videoconferencing works ☐
b Uses of videoconferencing ☐
c Guidelines for having a videoconference ☐
d Problems with bandwidth ☐

1 A videoconference lets people in different places see and hear each other at the same time. People use it for education, business and community events. Students can learn about different cultures in real time, and go on virtual field trips without leaving home. Businesses use it for meetings and job interviews because it saves money and time in travelling. Libraries and town halls can use it to bring people together for community meetings and other special activities.

2 Videoconferencing needs a Web camera and videoconferencing software. You can use the Internet, a Local Area Network (LAN) or an Integrated Service Digital Network (ISDN) to have a videoconference. A LAN is usually a closed network connected by wire cables. ISDN uses telephone lines but needs special adaptors instead of modems to send data.

3 Videoconferencing over the public Internet is not always reliable because the amount of data that you can send depends on bandwidth. Public telephone lines have a low bandwidth and usually give small video frames, poor picture quality and slow delivery. Broadband sends more information over the Internet at faster speeds but it is expensive.

4 Videoconferencing tips
• Keep your eyes on the Web cam to show you are interested.
• Move slowly and talk in a strong, clear voice because of the small time delay in videoconferencing.
• Wear dark or neutral colours as bright colours and patterns can affect picture quality.

3 Decide if the sentences are true (T) or false (F).

1 Businesses use videoconferencing for meetings. T/F
2 The high bandwidth on the public Internet delivers pictures faster. T/F
3 ISDN sends data using telephone lines. T/F
4 You should talk and move quickly in a videoconference. T/F
5 Grey, cream and light brown are good colours for videoconferences. T/F

4 Circle the method of videoconferencing which:

1 needs wire cables. LAN/ISDN
2 uses adaptors. LAN/ISDN
3 is usually a closed network. LAN/ISDN
4 does not use telephone lines. LAN/ISDN

Vocabulary

5 Complete the sentences (1–6) with the highlighted words in the text.

1 _____ let you use electronic devices anywhere in the world.
2 _____ is expensive but it is better for videoconferencing.
3 A low _____ gives you poor picture quality.
4 Broadband is more _____ than the public Internet.
5 With this software, you can make the video _____ larger.
6 The way you dress can _____ the video image people see.

Speaking

6 Work in groups. Imagine that you are going to have a videoconference with a school from another country.

• Think of how to introduce yourself.
• Make up a list of questions to ask the students about their country.
• Decide what to show and tell them about your school, country, customs and culture.
• Decide what to wear and how to move.
• Think of how to end your videoconference.

Take turns to practise your videoconference: one group asks questions and the group answers.

▶ ## Get real

Use the Internet to find advice on giving videoconference presentations. Find out about using visual displays, what to say when opening and closing videoconferences and how to deal with large numbers of people. Report back to the class.

Before you start

1 Who uses computers and the Internet more, boys or girls? Can you think of any reasons for this?

Reading

2 Read the text quickly. Write the correct heading above each section.

Why are boys better at IT?   Why is IT important?
Is the world changing?   Did you know that …

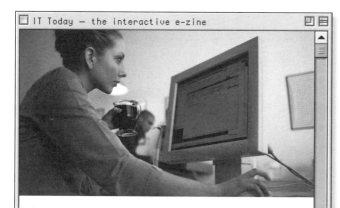

IT Today – the interactive e-zine

1 _____

• more men use IT than women?
• only 33% of the people studying IT are girls?
• only 4% of computer scientists are women?
This is strange because it's a fact that girls are just as clever as boys in science and mathematics.

2 _____

The usual explanation is that boys learn basic computer skills through video games. Girls do not usually like these violent and aggressive games, so boys have earlier experience with computers. What's more, when the Internet started, men did not encourage women to join. Many women who use the Internet complain that men are rude and unpleasant to them.

3 _____

• More than 75% of future jobs will need people with computer skills.
• Thousands of companies use the Internet to advertise job vacancies.
• Computers are tools, not just toys, and they can help everyone get good jobs.
You don't need to be a genius to learn computing. It just takes practice!

4 _____

Yes! 51% of new Internet users are women. More people are using Netiquette, which encourages women and girls to go online. More girls are learning computing, and programmers are designing imaginative and non-violent games that are fun and exciting to play.

3 Match the charts with the figures from the text that they illustrate.

1 Girls and boys studying IT ☐
2 New Internet users ☐
3 Computer scientists ☐

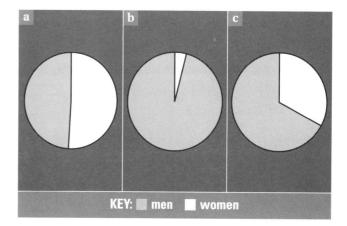

KEY: ■ men   ☐ women

4 Match the first part of the sentence (1–5) with second part (a–e).

1 If you want to learn computing,
2 Girls prefer playing
3 You can find a lot of jobs
4 It is a fact that
5 Computers are not just toys,

a girls are as good as boys at IT.
b by surfing the Internet.
c you must practise a lot.
d but they enable you to do many things.
e interesting and creative video games.

Vocabulary

5 Circle the word in each group that is different.

1	imaginative	creative	boring	interesting
2	help	encourage	support	stop
3	fact	opinion	belief	idea
4	typical	strange	unusual	unlike
5	aggressive	rude	unpleasant	nice
6	violent	peaceful	gentle	helpful
7	silly	genius	intelligent	clever
8	complain	promise	guarantee	agree

Speaking

6 Describe some computer games that you like. Are they more for boys, girls or both? Say why.

▶ **Get real**

Girls Find and play a video game that you think is more for boys.
Boys Find and play a video game that you think is more for girls.
Report back to the class, saying whether you liked the game or not, and why.

28 | Careers in IT

Before you start

1 Work in pairs and discuss the questions.

1 What are your plans when you finish school?
2 Do you want to use, or think you will need to use, IT in your job?
3 Make a list of jobs which use IT.

Reading

2 Read the quotes and write the name of the students by the jobs they want.

1 Web designer _____
2 Computer programmer _____
3 Database administrator _____
4 E-commerce manager _____

Elissa

"I'm interested in writing software. My friends say I'm a techno-nerd because I prefer working with computers to people. Money is important but I'd rather do a job I enjoy. I want to take a distance-learning course so I can study at home."

Katie

"I like shopping and I think the future of business is on the Internet. I'm good with computers but I also like working with people. I'd like to manage my own online company. This will give me a lot of responsibility. E-commerce comes with risks, but the rewards are high when you succeed."

Martin

"Many people like Web design, but I think data management gives more job security. There is so much information on the Internet, and companies need people who know how to store, manage and retrieve data. I want to get my degree and work for a good company."

Peter

"I'm using JavaScript to make my website more interactive. After college, I'd like to try telecommuting. This is working at home, using e-mail to communicate with clients. I want freedom, flexibility and long holidays, which you don't get by working in an office."

3 Write E, K, M or P. Which student:

1 wants to work at home?
2 wants a secure job?
3 does not want to study in college?
4 wants to choose when to work?
5 wants to manage people?
6 likes working with data?
7 wants to be rich and successful?
8 uses a coding system for web pages?

Vocabulary

4 <u>Underline</u> the ways of expressing *like* or *want* in the quotes, then choose the correct answer.

1 _____ telecommuting to working in an office.
  a I'd rather   b I prefer   c I like

2 _____ to do a distance-learning course.
  a I'd prefer   b I'd rather   c I don't like

3 _____ working long hours all the time.
  a I'd prefer   b I don't like   c It's good

4 _____ to work with computers all day as I think it would be boring.
  a I'd rather not   b I wouldn't like   c I don't like

5 _____ be a rich techno-nerd than poor and popular.
  a I'd rather   b I prefer   c I like

6 _____ in being a secretary. I want a better job.
  a I'm not interested   b I'm thinking of
  c I don't like

Speaking

5 Work in groups. Rank the things you want from a job: 1 = most important, 10 = least important.

- a high salary ■ flexible working hours
- ■ responsibility ■ interest or enjoyment
- ■ a nice office ■ telecommuting
- ■ long holidays ■ working with people
- ■ security ■ excitement/risk
- ■ good benefits, e.g. a company car, gym membership

Writing

6 Write a paragraph to say what kind of job you would like, and why.

Get real

Choose an area of IT that you are interested in. Find information about courses offered by colleges and universities. Find other areas where there are jobs in IT. Report back to the class on what you need to start the course or to get a good job.

Glossary

Note: Definitions taken or adapted from the Oxford Student's Dictionary © Oxford University Press. ISBN 0 19 431517 7.

Short forms

[C]	countable	*adv*	adverb
[U]	uncountable	*prep*	preposition
[pl]	plural	*sb*	somebody
adj	adjective	*sth*	something

A

access /'ækses/ *verb* to be able to open a file, website, program, database, etc.

accessory /ək'sesəri/ *noun* [C] an extra item that is added to sth and is useful or attractive but important

according to /ə'kɔːdɪŋ tə; before vowels tu/ *prep* in a way that matches, follows or depends on sth

adaptor /ə'dæptə/ *noun* [C] a device for connecting pieces of electrical equipment that were not designed to be fitted together

adjust /ə'dʒʌst/ *verb* to change sth slightly

affect /ə'fekt/ *verb* to make sb/sth change in a particular way; to influence sb/sth

algorithm /'ælgərɪðəm/ *noun* [C] a set of steps or instructions for solving a particular problem

align /ə'lain/ *verb* to arrange things in a straight line

animation /ˌænɪ'meɪʃn/ *noun* [C] a moving cartoon

antenna /æn'tenə/ *noun* [C] a piece of equipment on a mobile phone, etc. that is used for sending and receiving communications signals

appear /ə'pɪə/ *verb* to come into view so that you can see sb/sth

application /ˌæplɪ'keɪʃn/ *noun* [C] a program that is designed to do a particular job

ASCII /'æski/ *abbr* **American Standard Code for Information Interchange**; a code used to represent English characters as numbers so that data can be moved between computers that use different programs

attachment /ə'tætʃmənt/ *noun* [C] a document or file that you send to sb using e-mail

audio /'ɔːdiəʊ/ *adj* connected with the recording of sound

automatically /ˌɔːtə'mætɪkli/ *adv* happening by itself, without needing a person to operate any controls

B

background /'bækgraʊnd/ *noun* [C] the picture or colour on the first screen that appears when you turn on the computer (= **the desktop**)

back-up /'bækʌp/ *noun* [U,C] a copy of a computer file, etc. that can be used if the original is lost or damaged

➤ **back sth up** *phrasal verb* to make a copy of a computer file, program, etc. that can be used in case the main one fails or needs extra support

band mode /'bænd məʊd/ *noun* [C] a control or setting on a mobile phone which can be set at 'digital' or 'dual'

bandwidth /'bændwɪdθ/ *noun* [C,U] the amount of data that can be sent or received in a fixed amount of time by a communications channel, for example wires or radio waves. The higher the bandwidth, the faster the communications channel can transmit the data.

bill /bɪl/ *noun* [C] a piece of paper that shows you how much money you owe sb for goods or services

binary /'baɪnəri/ *adj* using only 0 and 1 as a system of numbers; *the binary system*

blink /blɪŋk/ *verb* (used about a light, text etc.) to come on and go off again quickly

Broadband /'brɔːdbænd/ *noun* [U] a communications medium that allows several channels of information, such as cable TV and Internet access, to pass through a single cable at the same time

browse /braʊz/ *verb* to look for or look at information on the Internet

browser /'braʊzə/ (also **Web browser**) *noun* [C] a program that lets you look at documents on the Internet

button /'bʌtn/ *noun* [C] (on computer screens) a small box that a user clicks, using a mouse, to tell the computer to do sth. A mouse also has left and right buttons.

C

carbon copy /ˌkɑːbən 'kɒpi/ *noun* [C] (abbr **cc**) a copy of a letter, an e-mail, etc. that is sent to sb else as well as the person it is addressed to

categorize /'kætəgəraɪz/ *verb* to divide people or things into groups

CD-rewriter /ˌsiː ˌdi ˌriː'raɪtə/ *noun* [C] (abbr **CD-RW**) a device that can be attached to a computer and that can read and write to writable disks

CD-ROM /ˌsiː diː 'rɒm/ *abbr* **compact disc read-only memory**; a CD, for use on a computer, which has data recorded on it. The data cannot be changed or removed, unlike **CD-RW** disks, on which data can be changed.

character /'kærəktə/ *noun* [C] a single letter, number or space that is typed in a computer document

chat room /'tʃæt rʊm/ *noun* [C] a **virtual** room on the Internet where people can communicate with each other

click /klɪk/ *verb* **click (sth/ on sth)** to press one of the buttons on a mouse to start an action on screen: *to click (on) a button/a hyperlink*

client /'klaɪənt/ *noun* [C] **1** a computer that is linked to a **server** and so can connect to a network to ask for files such as e-mail messages, Web pages and programs, and can also access stored data **2** a person who uses the services or advice of a professional person or an organization

clip art /'klɪp ɑːt/ *noun* [U] electronic images which you can download from the Internet or copy from **CD-ROMs**

clipboard /'klɪpbɔːd/ *noun* [C] a file or memory area where computer data is kept for a short time while the user cuts or copies sth from an open document

code /kəʊd/ *noun* [U] a set of written computer instructions

coding system /'kəʊdɪŋ sɪstəm/ *noun* [C] a way of representing data

combine /kəm'baɪn/ *verb* to join or mix two or more things together

command /kə'mɑːnd/ *noun* [C] an instruction that tells a computer what to do

commerce /'kɒmɜːs/ *noun* [C] the business of buying and selling things

communication /kəˌmjuːnɪ'keɪʃn/ *noun* [U] the act of sharing or exchanging information, ideas or feelings

compatible /kəm'pætəbl/ *adj* able to be used together

component /kəm'pəʊnənt/ *noun* [C] one of several parts of which a machine, etc. is made

compose /kəm'pəʊz/ *verb* to produce a piece of writing, etc.

computer programmer /kəmˌpjuːtə 'prəʊgræmə/ *noun* [C] a person whose job is to write programs for a computer

confident /'kɒnfɪdənt/ *adj* feeling or showing that you are sure about your own abilities, opinions, etc.

confusing /kən'fjuːzɪŋ/ *adj* difficult to understand; not clear

connection /kə'nekʃn/ *noun* [C,U] a point, especially in an electrical system, where two parts connect; the state of being connected together

consistent /kən'sɪstənt/ *adj* not changing; always having the same opinions, standards, etc.

consumer /kən'sjuːmə/ *noun* [C] a person who buys things or uses services

convenient /kən'viːniənt/ *adj* suitable or practical for a particular purpose; not causing difficulty

copyright law /'kɒpiraɪt lɔː/ *noun* [C,U] a law which gives sb the legal right to print, copy, etc. a piece of original work

counter /'kaʊntə/ *noun* [C] an electronic device or computer program for counting sth

CPU /ˌsiː piː 'juː/ *abbr* **central processing unit**; the part of a computer that controls all the other parts of the system, such as memory, speed and power supply

crash /kræʃ/ *verb* (used about a computer) to suddenly stop working

cross-platform /ˌkrɒs 'plætfɔːm/ *adj* (used about programs or hardware) that can be used in different **operating systems**

cursor /'kɜːsə/ *noun* [C] the small flashing mark on a computer screen that shows where the next **character** (= letter or number) on the screen will be displayed

custom /'kʌstəm/ *noun* [C,U] a way of behaving which a particular group or society has had for a long time

cut /kʌt/ *verb* to remove sth or part of sth on a computer screen

Information Technology **35**

cyberspace /'saɪbəspeɪs/ *noun* [U] the **virtual** place where electronic messages, pictures, etc. exist while they are being sent between computers

D

data /'dɑːtə/ *noun* [U] information that is stored by a computer

database /'deɪtəbeɪs/ *noun* [C] a collection of data organized in a way that allows you to **access**, **retrieve** and use it

database administrator /'deɪtəbeɪs əd,mɪnɪstreɪtə/ *noun* [C] a person whose job is to design and manage a database

deal with sth *phrasal verb* to carry out a task or take care of sth

decrease /dɪ'kriːs/ *verb* to make sth smaller or less

default /dɪ'fɔːlt; 'diː/ *noun* [U, C] what usually happens or appears on a computer screen if you do not make any other choice or change: *You can change the **default** settings* ➤ **default** *verb*

delete /dɪ'liːt/ *verb* to remove sth that has been stored on a computer

design /dɪ'zaɪn/ *noun* [C,U] (the process or skill of making) a drawing or plan that shows how sth new will be made, how it will work, etc.

desktop /'desktɒp/ *noun* [C] the first screen that appears when you turn on your computer and which displays icons that represent files, folders, documents, etc.

desktop publishing /,desktɒp 'pʌblɪʃɪŋ/ (*abbr* **DTP** /,diː tiː 'piː/) *noun* [U] using a personal computer to produce books, magazines, etc.

detachable /dɪ'tætʃəbl/ *adj* that can be taken off

develop /dɪ'veləp/ *verb* to make sth grow slowly, increase or change into sth else

device /dɪ'vaɪs/ *noun* [C] a tool or a piece of equipment made for a particular purpose

dialog box /'daɪəlɒg bɒks/ *noun* [C] a box that appears on a computer screen asking you to choose what you want to do next by typing or clicking buttons

digital /'dɪdʒɪtl/ *adj* using an electronic system that uses the numbers 1 and 0 to store **data**: *a digital camera*

dimension /dɪ'menʃn/ *noun* [C] a measurement in space, for example the height, width or length of sth

directory /də'rektəri/ *noun* [C] a list of the files or programs stored on a computer's hard drive

Display /dɪ'spleɪ/ *noun* [sing] a particular feature of **Windows**® that lets you change the way your computer screen looks by choosing your own background, screen saver, etc.

display screen /dɪ'spleɪ skriːn/ *noun* [C] the window where words, pictures, etc. are shown on a mobile phone

distance learning /,dɪstəns 'lɜːnɪŋ/ *noun* [U] the act of studying a subject or preparing for an exam from your home, away from a school or university

DNS /,diː en 'es/ *abbr* **Domain Name Service**; an Internet service that reads an e-mail address and translates it into a number (**the IP address**) that matches the e-mail address

domain name /də'meɪn neɪm/ *noun* [C] an **IP** (= **Internet Protocol**) address, written using text. It must have at least two parts, for example 'hotmail.com'.

double-click /,dʌbl 'klɪk/ *verb* (**double-click sth/on sth**) to press one of the buttons on a mouse twice quickly in order to start an action on screen: *double-click (on) a button/hyperlink*

download /,daʊn'ləʊd/ *verb* to copy data, such as a file, Web page or computer program from the Internet onto your computer

draft /drɑːft/ *noun* [C] a piece of writing, etc. which will probably be changed and improved before the final version

drive /draɪv/ *noun* [C] the part of the computer that reads and stores information on disks

drop-down menu /,drɒp daʊn 'menjuː/ *noun* [C] a list of possible choices that appears on a computer screen when you click on a title at the top

dual /'djuːəl/ *adj* having or using two parts or aspects

E

e-commerce /'iːkɒmɜːs/ *noun* [U] the buying and selling of goods and services on the Internet

edit /'edɪt/ *verb* to prepare a piece of text to be published, making sure that it is correct, the right length, etc.

else /els/ *adv* (used after words formed with any-, no-, some-, and after question words) another, different person, thing or place

e-mail (also **email**) /'iːmeɪl/ *noun* [C,U] **1** [U] a way of sending electronic messages or data from one computer to another **2** [C,U] a message or messages sent by e-mail ➤ **e-mail** *verb*

emoticon /ɪ'məʊtɪkɒn/ *noun* [C] a group of keyboard symbols that represent the expression on sb's face, used in e-mail, etc. to show the feelings of the person sending the message, for example :-) represents a smiling face

empty /'empti/ *adj* with nothing inside

EPS /,iː piː 'es/ *abbr* **Encapsulated Post Script**; EPS is part of the **Adobe® Systems** file format that you can use with most **desktop publishing** and **image editing programs**

erase /ɪ'reɪz/ *verb* to remove all or part of an image in an **image editing program**

exclude /ɪk'skluːd/ *verb* to leave out or not include sb/sth in sth

existing /ɪg'zɪstɪŋ/ *adj* that is already there or being used; present

export /ɪk'spɔːt/ *verb* to **format** data so that it can be used by another **application**

e-zine /'iː ziːn/ *noun* [C] an interactive magazine on the Internet

F

faceplate /'feɪspleɪt/ *noun* [C] the front part of a mobile phone

FAQ /,ef eɪ 'kjuː/ *abbr* (used in writing) **frequently asked question(s)**

feature /'fiːtʃə/ *noun* [C] something important, interesting or typical of a place or thing

fee /fiː/ *noun* [C] the money you pay for a service or for professional advice

field /fiːld/ *noun* [C] a text box where you type in words or data

file /faɪl/ *noun* [C] a collection of information, such as a Word document or a picture, which is stored in a computer, under a particular name

file extension /'faɪl ɪk,stenʃn/ *noun* [C] the last part of a file name, which shows you the **format** of the file, for example *.avi* for video, *.doc* for MS Word files, etc.

fit /fɪt/ *verb* to be the right size or shape for sb/sth

flame /fleɪm/ *noun* [C] (slang) a hurtful or insulting message that is sent to sb on the Internet ➤ **flame** *verb*

Flash™ /flæʃ/ *noun* [U] an interactive animation technology developed by **Macromedia Inc**

flip cover /'flɪp kʌvə/ *noun* [C] a cover for a mobile phone that you can open or close by turning it over

floppy disk /,flɒpi 'dɪsk/ *noun* [C] a flat disk inside a plastic cover, that is used to store information (**data**) in a form that a computer can read, and that can be removed from the computer

folder /'fəʊldə/ *noun* [C] a place where a number of computer files or documents can be stored together

font /fɒnt/ *noun* [C] the particular style of a set of letters that are used in printing, etc., such as *Times New Roman*

format /'fɔːmæt/ *verb* to change or arrange text in a particular way on a page or screen

frame /freɪm/ *noun* [C] **1** (in **videoconferencing**) a single image in a video clip **2** (in **HTML**) a box which divides a browser into different sections. Each **frame** is a different **Web page**. **3** (in **DTP applications**) a box containing text or pictures

freeware /'friːweə/ *noun* [U] software which you can download free from the Internet, without having to pay for it

FTP /,ef tiː 'piː/ *abbr* **File Transfer Protocol**; a set of rules that lets you move files from one place to another over a network. An **FTP** server is the computer that uploads and downloads files.

function /'fʌŋkʃn/ *noun* [C] the purpose or special activity of sth/sb

G

generate /'dʒenəreɪt/ *verb* to create or produce sth

GIF /ɡɪf/ *abbr* **Graphics Interchange Format**; a file format that is good for pictures or images that only use a few colours

gigabyte /ˈɡɪɡəbaɪt/ *noun* [C] (*abbr* **GB**) a unit of measurement used to measure the size of the hard disk. 1 **gigabyte** = 1,024 **megabytes**.

gigahertz /ˈɡɪɡəhɜːts/ *noun* [C] (*abbr* **GHz**) a unit for measuring the speed of a **CPU**. One GHz represents one billion cycles per second.

graphics /ˈɡræfɪks/ *noun* [pl] pictures or images that are used especially in the design of magazines, Web pages, etc.

guarantee /ˌɡærənˈtiː/ *noun* [C] a firm promise that sth will be done or that sth will happen

H

hard disk /ˌhɑːd ˈdɪsk/ *noun* [C] a disk inside a computer that stores all the data and programs in the computer

hardware /ˈhɑːdweə/ *noun* [U] the machinery and electronic parts of a computer system that you can touch, such as the keyboard, the **CPU**, etc.

high level language /ˌhaɪ levl ˈlæŋɡwɪdʒ/ *noun* [C] a programming language which is closer to human language than low-level computer languages, such as **machine language**

highlight /ˈhaɪlaɪt/ *verb* to mark part of a text with a special coloured pen, or to mark an area on a computer screen to emphasize it or make it easier to see

home page /ˈhəʊm peɪdʒ/ *noun* [C] the first of a number of pages of information on the Internet that belongs to a person or an organization. A **home page** contains connections (**links**) to other pages of information.

HTML /ˌeɪtʃ tiː em ˈel/ *abbr* **Hyper Text Mark up Language**; a system (a **Web authoring language**) used to create documents for the **World Wide Web**

HTTP /ˌeɪtʃ tiː tiː ˈpiː/ *abbr* **Hyper Text Transfer Protocol**; the **protocol** (= rules) used to send and receive data on the **World Wide Web**

hyperlink /ˈhaɪpəlɪŋk/ *noun* [C] a connection to a **Web page** or part of a **Web page**

I

icon /ˈaɪkɒn/ *noun* [C] a small symbol on a computer screen which represents a program, or a file

illegal /ɪˈliːɡl/ *adj* against the law

image /ˈɪmɪdʒ/ *noun* [C] a copy or picture of sth seen on a computer

import /ɪmˈpɔːt/ *verb* to use data produced by another **application**

income /ˈɪnkəm/ *noun* [C,U] the money that you receive regularly as payment for your work

incoming /ˈɪnkʌmɪŋ/ *adj* arriving somewhere, or being received

increase /ɪnˈkriːs/ *verb* to make sth bigger or greater

information technology /ˌɪnfəˈmeɪʃn tekˌnɒlədʒi/ (*abbr* **IT** /ˌaɪ ˈtiː/) *noun* [U] the study or use of electronic equipment, especially computers, for collecting, storing and sending out information

input /ˈɪnpʊt/ *noun* [U] the act of putting information into a computer

insert /ɪnˈsɜːt/ *verb* to put sth into sth or between two things

instant /ˈɪnstənt/ *adj* happening immediately

instead of /ɪnˈsted əv/ *adv, prep* in the place of sb/sth

integrate /ˈɪntɪɡreɪt/ *verb* to join things together so that they become one thing or work together

interact /ˌɪntərˈækt/ *verb* (used about a computer system and its user) to communicate directly with each other
➤ **interaction** /ˌɪntərˈækʃn/ *noun* [U,C]

interactive /ˌɪntərˈæktɪv/ *adj* involving direct communication between a computer and the person using it

Internet /ˈɪntənet/ *noun* [sing] (**the Internet**) a worldwide network that connects millions of computers

Internet Protocol Address /ˌɪntənet ˈprəʊtəkɒl ədres/ (also **IP address**) *noun* [C] a number used to identify a computer or device on a network

invasion /ɪnˈveɪʒn/ *noun* [C] the action of entering a place where you are not wanted and disturbing sb: *Such actions are an invasion of privacy.*

ISDN /ˌaɪ es ˌdiː ˈen/ *abbr* **Integrated Services Digital Network**; an international communications standard for sending data over digital telephone lines

J

JavaScript /ˈdʒɑːvəskrɪpt/ *noun* [U] a simple programming language that allows **Web authors** to design interactive **Web pages**

JPEG /ˈdʒeɪpeɡ/ *abbr* **Joint Photographic Expert Group**; a file type for storing photographs and images

junk mail /ˈdʒʌŋk meɪl/ *noun* [U] advertisements, etc. sent by post to people who have not asked for them

K

keyboard /ˈkiːbɔːd/ *noun* [C] the set of buttons (**keys**) that you press to operate a computer

keypad /ˈkiːpæd/ *noun* [C] a very small keyboard or set of buttons used for operating a small electronic device such as a mobile phone

keyword /ˈkiːwɜːd/ *noun* [C] a word or phrase that you type in when using a **search engine** to look for information on the Internet

L

LAN /læn/ *abbr* **local area network**; a network of computers within a single building or group of nearby buildings

link /lɪŋk/ *verb* to make a connection between two or more people or things

locate /ləʊˈkeɪt/ *verb* to find the exact position of sb/sth

location /ləʊˈkeɪʃn/ *noun* [C] a place where sth happens or exists

logical operator /ˌlɒdʒɪkl ˈɒpəreɪtə/ *noun* [C] a word (such as *and, or, not*) that is used in programming languages, when using **search engines**, etc. to give a computer more exact instructions about what it should look for or do

M

machine language /məˈʃiːn læŋɡwɪdʒ/ *noun* [C] a low-level computer language that is only made up of 1s and 0s. It is the only language that a computer understands.

mail server /ˈmeɪl sɜːvə/ *noun* [C] a computer and/or software that runs an e-mail system

main /meɪn/ *adj* most important

manual /ˈmænjuəl/ *adj* done or controlled by hand rather than automatically

maximize /ˈmæksɪmaɪz/ *verb* to make one window on a computer screen bigger in size, so that it covers the whole screen

means /miːnz/ *noun* [C] a method of doing sth

megabyte /ˈmeɡəbaɪt/ *noun* [C] (*abbr* **MB**) a unit for measuring computer memory. 1 megabyte = 1,048,576 bytes. 1 byte = 1 single typed letter, number or space (**character**).

megahertz /ˈmeɡəhɜːts/ *noun* [C] (*pl* **megahertz**) (*abbr* **MHz**) a unit for measuring the speed of a **CPU**. One MHz represents one million cycles per second.

menu bar /ˈmenju bɑː/ *noun* [C] a row of words or commands (**File**, **Edit**, etc.) that are shown at the top of a computer screen

minimize /ˈmɪnɪmaɪz/ *verb* to make a window on a computer into an **icon**

mobile phone /ˌməʊbaɪl ˈfəʊn/ (also **'mobile**) *noun* [C] a small telephone without any wires that works by radio and that you can carry around with you

modem /ˈməʊdem/ *noun* [C] a device that connects a computer to the Internet. It changes computer data into sound which can be sent over telephone lines.

monitor /ˈmɒnɪtə/ *noun* [C] a separate part of a **PC** with a large screen that shows information from the computer

mouse /maʊs/ *noun* [C] a small device that you move across a surface with your hand to control the movement of the **cursor**

multilingual /ˌmʌltiˈlɪŋɡwəl/ *adj* for or including people of many different languages and races

multimedia /ˌmʌltiˈmiːdiə/ *adj* using sound, pictures and video in addition to text on a screen

multiple /ˈmʌltɪpl/ *adj* involving more than one person or thing; having many parts

N

narrow /ˈnærəʊ/ *verb* to make sth smaller or less wide

navigate /'nævɪgeɪt/ *verb* to use a map, etc. to move around a place or find your way somewhere

navigation /ˌnævɪ'geɪʃn/ *noun* [U] the ability to find your way easily around somewhere/something

navigation bar /ˌnævɪ'geɪʃn bɑː/ *noun* [C] the list of words or images at the top, bottom or side of a **home page** that shows you where to find everything on a **website**

network /'netwɜːk/ *noun* [C] a number of computers and other devices that are connected together so that equipment and information can be shared

notebook /'nəʊtbʊk/ *noun* [C] (also **laptop** /'læptɒp/) a small personal computer that you can carry

O

offline /ˌɒf'laɪn/ *adj, adv* not connected to the Internet

online /ˌɒn'laɪn/ *adj, adv* connected to the Internet

online community /ˌɒnˌlaɪn kə'mjuːnəti/ *noun* [C] a group of people who all have sth in common and who meet and communicate regularly on the Internet

operate /'ɒpəreɪt/ *verb* to manage or use sth; to do business

optional /'ɒpʃənl/ *adj* that you can choose to do or not do

organize /'ɔːgənaɪz/ *verb* to put sth in order; to tidy sth

original /ə'rɪdʒənl/ *adj* new and interesting; different from its type

originate /ə'rɪdʒɪneɪt/ *verb* to happen or appear for the first time in a particular place or situation

outgoing /'aʊtgəʊɪŋ/ *adj* going away from a particular place, or being sent

output /'aʊtpʊt/ *noun* [U,C] the information that a computer produces

P

paint /peɪnt/ *verb* to electronically fill an area with colour using an **image editing tool**

password /'pɑːswɜːd/ *noun* [C] a secret word or series of numbers that you type into a **text box** in order to use a program or a computer

paste /peɪst/ *verb* to copy or move text or graphics into a document from somewhere else

PC /ˌpiː 'siː/ *abbr* **personal computer**; the general term used for a computer, which usually consists of a monitor, a tower, a keyboard and a mouse

peer-to-peer /ˌpɪə tə 'pɪə/ *adj* from one individual computer user to another, without using a central **server**

performance /pə'fɔːməns/ *noun* [U] how well or badly a computer, machine, etc. works or does sth

peripheral /pə'rɪfərəl/ *noun* [C] any piece of **hardware** (= machinery, etc. that forms part of or connects to a computer) apart from the **CPU** and the working memory (**RAM**)

personal information /ˌpɜːsənl ˌɪnfə'meɪʃn/ *noun* [U] details about yourself; where you live, your phone number, etc.

personalize /'pɜːsənəlaɪz/ *verb* to design or change sth so that it is suitable for the needs of one particular person

PICT *abbr* a file format for **graphics**, designed for **Apple Macintosh** computers

plug-in /'plʌgɪn/ *noun* [C] a program, a file or **hardware component** that adds a specific feature to a computer system

pointer /'pɔɪntə/ *noun* [C] a small arrow on a computer screen that you move by moving the mouse

pop-up ad /ˌpɒp ʌp 'æd/ *noun* [C] an advertisement that suddenly appears when you are surfing the Internet

printer /'prɪntə/ *noun* [C] a machine that is connected to a computer and that prints on paper

privacy /'prɪvəsi/ *noun* [U] the state of not being seen or disturbed by other people

privacy policy /'prɪvəsi pɒləsi/ *noun* [C] a plan of action agreed by a company, which promises that any personal information you give to that company will not be passed on to anyone else

procedure /prə'siːdʒə/ *noun* [C, U] a way of doing sth

process /'prəʊses/ *verb* (used about a computer) to read data and use it to perform a series of tasks (**operations**)

program /'prəʊgræm/ *verb* to give a set of instructions to a computer to make it perform a particular task

protocol /'prəʊtəkɒl/ *noun* [C] a system of rules for transmitting data between two devices, for example **Post Office Protocol**

R

RAM /ræm/ *abbr* **random access memory**; computer memory in which data can be changed or removed and then looked at in any order. When you switch the computer off, you lose all the data in the **RAM**.

random /'rændəm/ *adj* done, chosen, etc. without sb thinking or deciding in advance what is going to happen

real time /ˌriːəl 'taɪm/ *noun* [U] the fact that there is only a very short time between a computer system receiving information and dealing with it, so that things almost happen live

recipient /rɪ'sɪpiənt/ *noun* [C] a person who receives sth

Recycle Bin /ˌriː'saɪkl bɪn/ *noun* [sing] the folder in **Microsoft Windows®** where files or programs that have been deleted or removed are stored

register /'redʒɪstə/ *verb* to put your name on an official list to ask for or join sth

related /rɪ'leɪtɪd/ *adj* connected with sb/sth

relevant /'reləvənt/ *adj* connected with what is happening or being talked about

reliable /rɪ'laɪəbl/ *adj* that you can trust

removable disk /rɪˌmuːvəbl 'dɪsk/ *noun* [C] a disk that stores data and that can be taken out of the computer and carried about with you

restore /rɪ'stɔː/ *verb* to put sth back into the position or condition it was in before

retailer /'riːteɪlə/ *noun* [C] a person or company who sells goods to the public

retrieve /rɪ'triːv/ *verb* (used about a computer) to find information that has been stored

ring tone /'rɪŋ təʊn/ *noun* [C] the sound or music that you program your mobile phone to play in order to tell you that sb is phoning you

rotate /rəʊ'teɪt/ *verb* to make sth turn in a circle around a central point

run /rʌn/ *verb* to use a computer program

S

save /seɪv/ *verb* to make a computer store and keep data

Save as *type noun* [U] the instruction or command that tells a computer how (= in what **file format**) you want it to store a particular file

Save in *noun* [U] the instruction or command that tells a computer where (= on which **drive**, in which **folder**, etc.) you want it to store or keep a file

scanner /'skænə/ *noun* [C] a piece of computer hardware that can read images on paper and change the information into data that a computer can use

screen saver /'skriːn seɪvə/ *noun* [C] a program that runs a moving image on a computer screen when the keyboard and the mouse are not being used

scroll bar /'skrəʊl bɑː/ *noun* [C] a tool on a computer screen that you use to move the text up and down or left and right

scroll key /'skrəʊl kiː/ *noun* [C] one of the buttons (**keys**) on a mobile phone which you press in order to move up or down inside the display screen

search /sɜːtʃ/ *verb* to examine sth carefully because you are looking for sth ▶ **search** *noun* [C]

search engine /'sɜːtʃ endʒɪn/ *noun* [C] a program (such as *Google™* or *Alta Vista™*) that lets you search the **World Wide Web** for information

secure /sɪ'kjʊə/ *adj* safe; well protected or locked

security /sɪ'kjʊərəti/ *noun* [U] the state of being safe and protected from danger, thieves, etc.

server /'sɜːvə/ *noun* [C] a central computer that lets people access information on a network and that stores data and programs centrally. There are many types of servers, such as **POP3 servers, network servers** and **Web servers**.

setting /'setɪŋ/ *noun* [C] one of the positions of the controls of a computer that control the way sth looks or works and that can be changed by the user: *paragraph settings*

share /ʃeə/ *verb* to divide sth between two or more people

shareware /'ʃeəweə/ *noun* [U] **software** which you can **download** from the Internet without having to pay for it until you have tried it for a time and decided that you want to continue to use it

shortcut /'ʃɔːtkʌt/ *noun* an **icon** that opens a program. A **shortcut key** is a keyboard operation that starts a command (for example pressing **Cntl** and **S** together to save a document). An **underscore** (e.g. S̲ave) shows a shortcut key in the **menu bar**. Shortcuts vary between computers and programs.

simulation /ˌsɪmjuˈleɪʃn/ *noun* [C,U] the creating of certain conditions that exist in real life using computers, etc., usually for study or training purposes

small talk /'smɔːl tɔːk/ *noun* [U] polite conversation, for example at a social event, about unimportant things

SMS /ˌes em ˈes/ *abbr* **Short Message Service** (or **Simple Message Service**); a service for sending text messages on mobile phones

software /'sɒftweə/ *noun* [U] the programs used by a computer

source /sɔːs/ *noun* [C] a place, person or thing where sth comes or starts from or where you can get sth

spam /spæm/ *noun* [U] advertising material, etc. that is sent by e-mail to people who have not asked for it

special effects /ˌspeʃl ɪˈfekts/ *noun* [pl] effects that can be created by computer graphics

specific /spəˈsɪfɪk/ *adj* clear and exact

specification /ˌspesɪfɪˈkeɪʃn/ *noun* [C] a detailed description of how sth is, or should be, designed or made

stand for sth *phrasal verb* to be an abbreviation or symbol of sth

stand-alone /'stænd ələʊn/ *adj* able to be operated on its own without being connected to a larger system

standard /'stændəd/ *adj* normal or average

store /stɔː/ *verb* to keep information or data in a computer's memory

structure /'strʌktʃə/ *verb* to plan or build sth in an organized way

stylish /'staɪlɪʃ/ *adj* fashionable and attractive

subject /'sʌbdʒɪkt/ *noun* [C] the topic or heading of an e-mail; the person or thing that is being talked about

support /səˈpɔːt/ *noun* [C] help and encouragement that you give to a person or thing

surf /sɜːf/ *verb* to look for or look at information on the Internet

swap /swɒp/ *verb* to give sth to sb else in exchange for sth

system /'sɪstəm/ *noun* [C] a particular way of doing sth

T

techno-nerd /'teknəʊnɜːd/ *noun* [C] a person who spends all his/her time on the Internet or working with new technology

template /'templeɪt/ *noun* [C] a plan of sth that is used as a model for producing other similar examples

text box /'tekst bɒks/ *noun* [C] a box that appears on a computer screen where a user can type in text or information

text editor /'tekst edɪtə/ *noun* [C] a program, such as *Notepad*, that allows you to write and edit text files

text message /'tekst mesɪdʒ/ *noun* [C] a written message that is sent from one mobile phone to another

text wrap /'tekst ræp/ *noun* [U] the act or process of arranging words around a picture or inside a shape in DTP programs

thesaurus /θɪˈsɔːrəs/ *noun* [C] a book that is like a dictionary, but in which words are arranged in groups that have similar meanings

3-D /ˌθriː ˈdiː/ *adj* having length, width and height

TIFF /tɪf/ *abbr* **Tagged Image File Format**; a file format for storing photographs and images

tool /tuːl/ *noun* [C] a thing that helps you to do your job or to achieve sth

toolbar /'tuːlbɑː/ *noun* [C] a row of symbols (**icons**), usually below the **menu bar**, that represent the different commands or tools that a user needs to use a program

tower /'taʊə/ *noun* [C] a metal box that contains the **CPU**, **hard disk drive** and power supply for a **PC**

transaction /trænˈzækʃn/ *noun* [C] a piece of business that is done between people

transfer /trænsˈfɜː/ *verb* to move sth from one place to another

translate /trænsˈleɪt/ *verb* to change sth written from one language into another

transmission /trænsˈmɪʃn/ *noun* [U] the action of sending sth out from one person, machine or thing to another

trial membership /ˌtraɪəl ˈmembəʃɪp/ *noun* [C] the state of being a member of a group, club, organization, etc. for a short period of time as a test, so that you can decide if you would like to continue permanently

U

underscore /ˌʌndəˈskɔː/ *verb* to underline sth ➤ **underscore** *noun* [C]

unsolicited /ˌʌnsəˈlɪsɪtɪd/ *adj* not asked for

untitled /ˌʌnˈtaɪtld/ *adj* having no title or name. A **graphics program** will usually store a file as 'untitled' if no other name is given.

upload /ˌʌpˈləʊd/ *verb* to copy a computer file from one computer system to another, usually on the Internet

URL /ˌjuː ɑːr ˈel/ *abbr* **Uniform/Universal Resource Locator**; the address of a **World Wide Web** page. URLs connect files across the **Web**.

utility /juːˈtɪləti/ *noun* [C] a program or part of a program that does a particular task or service

V

videoconferencing /'vɪdiəʊkɒnfərənsɪŋ/ *noun* [U] a system that allows people in different parts of the world to have a meeting by watching and listening to each other using video screens

view /vjuː/ *verb* to look at or be able to see sth

virtual /'vɜːtʃuəl/ *adj* made to appear to exist

virtual reality /ˌvɜːtʃuəl riˈæləti/ *noun* [U] images created by a computer that appear to surround the person looking at them and seem almost real

virus /'vaɪrəs/ *noun* [C] a computer program that attaches itself to another program in order to destroy files or damage the **hard disk** of the computer

voicemail /'vɔɪsmeɪl/ *noun* [U] an electronic system which can store telephone messages, so that sb can listen to them later

W

Web camera /'web kæmərə/ (also **Webcam; webcam** /'webkæm/) *noun* [C] a video camera connected to a computer that is connected to the Internet, so that its images can be seen by Internet users

Web design /'web dɪzaɪn/ *noun* [U] the job or activity of drawing or planning **Web pages**

Web page /'web peɪdʒ/ *noun* [C] a collection of text, pictures, sound etc. that you see in a window on your screen when you visit a **website**. A **website** consists of several **Web pages**.

Web-authoring /'web ɔːθərɪŋ/ *noun* [U] the process of creating **Web pages** by writing **HTML code** (= an authoring language)

Web-based /'web beɪst/ *adj* connected to, or made available via, the Internet and the **World Wide Web**

website /'websaɪt/ *noun* [C] a place connected to the Internet, where a company, an organization, etc. puts information that can be found on the **World Wide Web**

Wi-Fi /'waɪ faɪ/ *noun* [U] **Wireless Fidelity**; a wireless technology for computers

wire cable /ˌwaɪə ˈkeɪbl/ *noun* [C] a type of cable used to connect computers

word processor /'wɜːd ˌprəʊsesə/ *noun* [C] a program or computer that lets you carry out word-processing functions

(the) World Wide Web /ˌwɜːld waɪd ˈweb/ (also **the Web**) *noun* [sing] (*abbr* **WWW** /ˌdʌbljuː dʌbljuː ˈdʌbljuː/) a worldwide collection of electronic documents **formatted** in **HTML**

OXFORD
UNIVERSITY PRESS

Great Clarendon Street, Oxford OX2 6DP

Oxford University Press is a department of the University
of Oxford. It furthers the University's objective of excellence in
research, scholarship, and education by publishing worldwide in

Oxford New York

Auckland Bangkok Buenos Aires Cape Town Chennai
Dar es Salaam Delhi Hong Kong Istanbul Karachi Kolkata
Kuala Lumpur Madrid Melbourne Mexico City Mumbai
Nairobi São Paulo Shanghai Taipei Tokyo Toronto

Oxford and Oxford English are registered trade marks of
Oxford University Press in the UK and in certain other countries

ISBN 0 19 438826 3

Printed in Spain by Unigraf S.L.

Acknowledgements

The authors and publishers are very grateful to the teachers who
read and piloted material and gave invaluable feedback.

Geert Claeys, Pavla Čípová, Blanka Klimova, Hana Sedláková,
Alena Šteklova

**We would like to thank the following for their kind
permission to reproduce photographs and other copyright
material:**
Photodisc Royalty Free pp 10, 12 (Andrea), 19, 25 (boy, Xmas
trees, cat), 33, 34 (Elissa, Katie)
Corbis Royalty Free pp 12 (Lida, Jarek), 25 (girl), 34 (Martin,
Petr)

Illustrations by:
Jeremy Banx p 31
Peter Harper pp 2, 4, 13

All other technical illustrations by Keith Shaw

Original references for the illustrations on pages 13, 16, 19, 21,
25, 30 and 33 were provided by the author.